3/13/12
$24.99

Architectural Trim

© 2007 by Quarry Books

All rights reserved. No part of this book may be reproduced in any form without written permission of the copyright owners. All images in this book have been reproduced with the knowledge and prior consent of the artists concerned, and no responsibility is accepted by the producer, publisher, or printer for any infringement of copyright or otherwise, arising from the contents of this publication. Every effort has been made to ensure that credits accurately comply with information supplied. We apologize for any inaccuracies that may have occurred and will resolve inaccurate or missing information in a subsequent reprinting of the book.

First published in the United States of America by
Quarry Books, a member of
Quayside Publishing Group
100 Cummings Center
Suite 406-L
Beverly, MA 01915-6101
Telephone: (978) 282-9590
Fax: (978) 283-2742
www.quarrybooks.com

Library of Congress Cataloging-in-Publication Data
Berry, Nancy E.
    Architectural trim: ideas, inspiration, and practical advice for adding wainscoting,
    mantels, built-ins, baseboards, cornices, casings, and columns to your home / Nancy E.
    Berry.
        p.   cm. — (Home design details)
    ISBN-13: 978-1-59253-326-8
    ISBN-10: 1-59253-326-4
    1. Trim carpentry—Amateurs' manuals. 2. Dwellings—Remodeling—Amateurs' manuals. I. Title.
TH5695.B47 2007
694'.6—dc22   2006032182
                                            CIP

ISBN-13: 978-1-59253-326-8
ISBN-10: 1-59253-326-4

10 9 8 7 6 5 4 3 2

Design: Stephen Gleason Design

Cover images: Gordon Beall, (top left); Brian Vanden Brink, (top right); Brian Vanden Brink/Catalano Architects, (bottom left); Erik Johnson/Chadsworth's 1.800.Columns, (bottom right)

Illustrations by: Robert Leanna II

Glossary of Trim Terms excerpted from Wikipedia

Printed in China

HOME DESIGN DETAILS

Architectural Trim

Ideas, Inspiration, and Practical Advice for Adding Wainscoting, Mantels,
Built-Ins, Baseboards, Cornices, Casings, and Columns to Your Home

NANCY E. BERRY

BEVERLY MASSACHUSETTS

QUARRY BOOKS

CONTENTS

INTRODUCTION

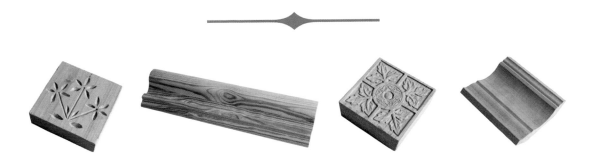

A room without trim simply looks naked.

Trim adds beauty, shadow, texture, and dimension to a space—not to mention concealing seams and uneven joints between walls, ceilings, and floors. Today, many new homes are lacking these subtle details, which add character and charm to interiors. In the housing industry, trim is all-too-often left out of the budget to be replaced by more unadorned square footage.

One of the things I love most about the 1870 Queen Anne in which I live is its rich detailing—especially the trim. This old Victorian's original 10" (25.4 cm)-baseboard and deep decorative crown moldings in the living room, dining room, and study add formality, beauty, and interest to these spaces. Even the simple flat-board door and window casings, along with the bead board wainscoting in the kitchen and pantry, add depth and character. These subtle details make the difference in a home's charm factor.

So when I decided to renovate an old barn on the property and turn it into my home office, I looked to the original trim in the house for ideas. I wanted to create that same charm found throughout the main house. Although I wasn't able to replicate the exact trim profiles, I came pretty close. By combining a few pieces of off-the-shelf trim from the local home center's lumberyard, I was able to create similar molding details to the 140-year-old finish carpentry work

found in the informal spaces of the main house. I also lined the hall leading to the renovated barn with 36" (91.4 cm) tongue-and-groove bead boards, similar to the kitchen's original Victorian wall paneling. The bead board covered a crumbling plaster surface while adding a fresh, casual look to the back entryway. I painted the new workroom a calming robin's egg blue and painted the trim a crisp white. With the bead board, baseboard, and window and door casings in place, the room became an inviting spot to work, read, and write.

Whether you live in an old house and are replicating existing trim for a sensitive addition, or you live in a new home where the trimwork is minimal—or even nonexistent—you have options for incorporating trim details into your room design. Architectural Trim covers all the steps for adding trim to your home, such as design inspiration to suit your personal style, how-to carpentry steps for adding elements like bead board or crown molding, information on appropriate period styles, and suggestions for decorative finishes to trim every room in your house. The book also highlights how you can create customized molding by using standard trim pieces available at home-improvement centers and lumberyards. Part One focuses on the trim basics: types, styles, materials, and finishes. Part Two provides the design inspiration for adding trim to your home through three-dimensional renderings of trim in a room.

PART ONE
Profile on Trim

Whether it is made of wood, fiber-glass, plastic, or plaster, architectural trim adds character and detail to your home. In an unadorned room, it adds depth and beauty to your design and truly finishes the space. In Part One of this book, we'll explore the basics of trim—from its history to the "how to" and every-thing in between.

Trim Basics

With the number of mundane production houses being built today, it's no wonder we're craving homes that have character and charm. Homes found in suburbia might have the square footage and acreage we want, but these new houses are often lacking in the subtleties that make them rich and warm. It's the design details that make the difference. Probably the most important architectural design element—and the one most often overlooked—is the home's interior trim. There was a time when trim was incorporated into both elegant mansions and simple farmhouses alike. Unfortunately, this practice has fallen by the wayside and today many production homes are assembled with minimal trim— often only a simple clamshell casing around the door and window openings. These new houses can look empty and sparse compared to homes built 100 years ago. How can we incorporate those trim details into our homes today? With a bit of ingenuity and planning, these empty canvases can be transformed into inviting, three-dimensional spaces.

Trim Talk

- Trim components consist of planes and curved (convex or concave) sections.

- Planes are flat moldings that reflect light in even tones while curved moldings reflect light in graded tones.

- Planes are broader and larger than curved moldings.

- Convex moldings accent with highlights while concave moldings cast shadows.

- Horizontal moldings offer balance and stability while vertical moldings and columns offer strength to a space.

Whether a simple flat board casing around a door or an elaborate Corinthian column screen, moldings give rooms personality and style.

Trim Design

Simply defined, trim is the finished woodwork or plaster-work in a room. The most basic types are baseboards, window and door trim, crown molding, and chair rails. When executed well, trim makes a room look and feel finished. The purpose of trim, also called moldings, is to create a play of light and shade within interior spaces. When sunlight rakes across trim, wonderful shadows and lines appear, adding beauty to a room. Trim, which is typically made up of two or more pieces, breaks up flat surfaces such as ceilings, walls, and floors, and defines the overall aesthetic in a room.

Some moldings are classified by their shape and others are classified by where they are placed in the room. For instance, a quarter round is named for its shape while a chair rail is named for its placement in the room.

Above: Many of today's trim profiles are based on ancient Greek and Roman architectural forms. The door trim is actually a fluted pilaster, a design from Greek architecture.

Right: Incorporating trim into your overall room design will add depth and dimension to the space. For example, this series of vaulted arches in a new home adds depth to this hallway as well as creates a frame for the front door.

Trim Origins

Today's ornamental trim is based on the classical orders of architecture developed in ancient Greece and Rome more than 2,000 years ago. An order is the arrangement of a particular style of column together with its entablature, which the column supports, and its standardized details, including its base and capital. The Greeks developed the Corinthian, Doric, and Ionic orders while the Romans added the Tuscan and Composite orders. Italian Renaissance architects of the sixteenth century based their building on these classical forms. Many trim profiles are derived from these ancient Greek and Roman columns and their entablatures. Eighteenth- and nineteenth-century architects and house builders looked to these ancient forms to create classical moldings in home design and these practices and profiles are still being used today. Some of the earliest types of trim are wainscoting or paneling, which are created by a succession of floor-to-ceiling chamfered vertical boards. These might have been used for insulation during the seventeen-century in cold climate areas. Another common trim component was the baseboard, which provided a covering between the floor and the wall intersection. Baseboards were often painted black during this time to disguise scuff marks. Chair rails were introduced to protect the wall from the backs of chairs, while picture rails and cornices form the uppermost horizontal trim in the room. These early moldings were cut with knives and hand planed. The carpenter often left his handcraft with subtle tool marks. By the time of the industrial revolution, mills started producing machined moldings and trim. Today, you can purchase trim at your local lumberyard or home center, or you can have a millshop custom make moldings for your home—and it is not as pricey as you might think.

Trim not only covers seams within a room but also adds punctuation to the room's elements. The crown signals that the wall is ending and the ceiling is beginning. The baseboard shows us the transition between the wall and the floor. The casings let you know that a door or window is beginning.

Trim Talk

- Millwork is the term for house components and trim made of wood that is sawed at a mill shop. These include door and window casings, porch posts, brackets, and window sashes.

- Moldings are decorative, linear bands and can be either wood or plaster.

Wainscoting topped with a plate rail adds a decorative and practical element to an Arts and Crafts–inspired dining room designed by architect Sandy Vitzthum. Note the simple bracket detailing under the shelf.

Wall Trim

Although home builders decked out walls with trim until about 1940, by mid-century, stripped walls became the norm. Even though home builders are seeing the value of incorporating trim into home design, many homeowners might want to build on existing trim or customize their own. In some cases, production homes are sold with little or no trim and the homeowner must start from scratch. Wall trim is a good place to start because it breaks up the expanse of larger rooms.

Baseboards

The most common type of wall trim is baseboard. Although a decorative feature, it also serves a practical purpose by covering the seam between the floor and the wall while protecting the lower wall from scuffs and bumps. Baseboard comes in a variety of profiles and sizes, and the more detailing the baseboard has, the richer it appears. In today's rooms with 8' (2.4 m) ceilings, baseboards are typically 6" to 8" (15.2 to 20.3 cm) high. Incorporating baseboards grounds the space and offers a finished look.

Chair Rails

Typically installed about 3' (0.9 m) above the floor, chair rails—like baseboards—are horizontal bands of trim. The chair rail is both decorative and functional, protecting the wall from the backs of chairs. A chair rail can stand alone atop a half-paneled wall, which is also known as wainscoting.

Wainscoting

Wainscoting is a type of wall paneling with a repetitive design and is used to cover walls from floor to ceiling. More decorative wainscoting has box moldings and chamfers etched into the wood's edges. Battens, which are narrow strips of wood, are also a good way to add a decorative touch to walls. Raised paneling is another traditional approach to adding a decorative component often used in studies and libraries. A structure of flat rectangular surfaces placed between stiles and rails is a popular option as well.

Bead board, a type of wainscoting, adds casual country charm to bathrooms, halls, and kitchens. A series of thin boards edged with a thin bead are placed vertically side by side to create a wall. Bead board can run up to a chair rail about one-third of the room's height to two-thirds up to what is called a plate rail. It is a great wall covering for kitchens, bathrooms, and mudrooms because it is much easier to clean than plain drywall.

Bead board wainscoting has been a decorative trim element in homes for centuries. Today, it can add dimension to otherwise flat walls.

Ceiling Trim

Ceiling trim can dramatically affect how large or small a room feels to its occupants. For example, adding a wide crown molding to a large room can make it feel cozy and intimate, and removing a low ceiling and exposing beams can transform a small bedroom into a dramatic one.

Crown Molding

Crown molding is the top horizontal strip of molding that covers the seam between the wall and the ceiling as well as offers visual interest and charm to a space. Its profile typically derives from classical architecture and can be as simple or dramatic as the design of the room allows. A popular crown molding choice is the cove molding. Its concave shape gives the illusion that the wall is curving toward the ceiling. Crown moldings can be made simply of just a few stock trim pieces or made more complex by using several custom pieces built up in layers—it depends on your personal style and budget.

Coved Ceilings

A coved ceiling, a concave surface or molding placed at the transition point between the wall and the ceiling, is another decorative alternative to a flat ceiling. This type of ceiling is often found in formal living rooms and dining rooms.

Beamed Ceilings

The use of beams is another decorative treatment for the ceiling. Depending on the design of your home, you can incorporate either a truss system for a rustic look or a box-beamed ceiling for a more formal look. Beams are either functional or purely aesthetic. Ceiling joists and trusses in timber-frame houses not only support the roof but also create a decorative accent within the structure. Many traditional architects search for salvaged or reclaimed joists and beams for new construction today to give the space an authentic, antique look. These salvaged components often can be old-growth wood, which has a stronger, tighter wood grain not found in today's harvested trees. These older pieces of wood add character—hand-adzed or planed, they show the marks of centuries-old craftsmanship on their well-worn surfaces.

Box-Beam Ceilings

Box-beam ceilings have a series of beams that run both vertically and horizontally across the ceiling in a house. They are made of lumber of various dimensions joined side by side to create a U shape. The beams are often adorned with molding pieces where the ends meet the ceiling. Hollow beams are the perfect place to hide wiring running though a room. They also can hide other unsightly construction necessities such as steel beams.

These decorative ceiling beams offer a rustic, country look. They also help visually break up the large expanse of the ceiling.

Door and Window Trim

Door and window trim and casings are an important and integral component to the overall design of a home. In ancient Greek architecture, door casings were more prominent than window casings, and this still holds true today. Casings hide the space between the wall and the door or window opening and create a handsome frame for the space beyond. Back bands—ornate molding features—can be added to the outside of casings for a more decorative look. Depending on your personal design tastes, casings can be elaborate, or simple, and represent every historical style. Typically, casings are made of soft or hard wood and are composed of both vertical and horizontal pieces. The casing can be as simple as a plain, flat board or as elaborate as a deep molding flanked with pilasters and topped with a Greek pediment. Casing seams can be disguised with corner blocks or rosettes at the top of the door.

Like all trim, window trim can be formal or plain but usually takes a backseat to the view beyond. Aprons, stools or sills, and stops are all components of window trim. When you are choosing window trim, the casing and the apron should have the same profiles for continuity, if not, the same scale.

Today, production houses come equipped with simple clamshell casings, which might be fine for upper floors or utilitarian spaces in your home, but if you want to add style and flair, you might consider outfitting your doors and windows with a more pronounced casing. Remember to keep casing consistent around door frames and window openings in adjoining rooms. This will create a sense of cohesiveness throughout the design.

An arch, a curved construction that spans an opening, often is reserved for formal living rooms and hallways within the home. These distinctive openings also receive trim. A more complicated building component, arches come in a variety of shapes and sizes. Wood trim around archways is often prefabricated to pre-existing dimensions or custom built by a skilled carpenter. There is also a synthetic product that can be easily curved to create the molding around the arch.

Trade Tip

Moldings are available in several standard types and sizes at your local home center. Moldings typically come in standard lengths of 8, 10, 12, 14, and 16 feet (2.4, 3, 3.7, 4.3, and 4.9 meters). Standard patterns are typically made of soft woods. If you're looking for more elaborate molding profiles, a local millwork shop can custom make moldings for you.

Window trim is often less ornate than door trim because window trim acts as a frame for the view beyond. Note that the casings and apron have the same profiles.

Architect Ben Nutter worked this plate rail into the casing above a bank of windows. Note the band of stock crown molding above the rail.

Trim Profiles

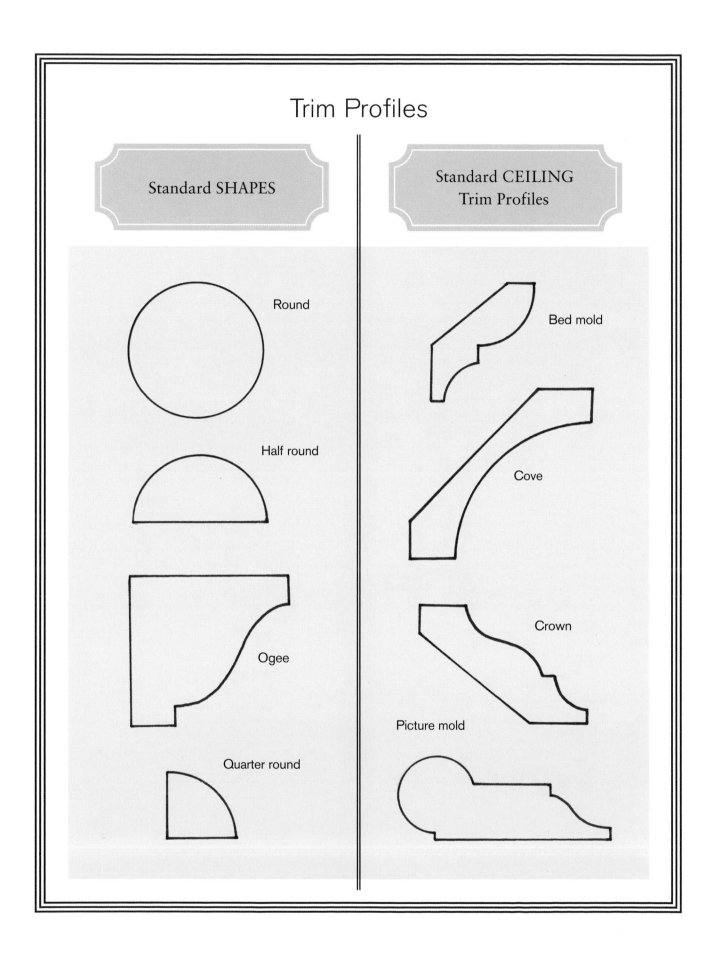

Standard SHAPES

Round

Half round

Ogee

Quarter round

Standard CEILING Trim Profiles

Bed mold

Cove

Crown

Picture mold

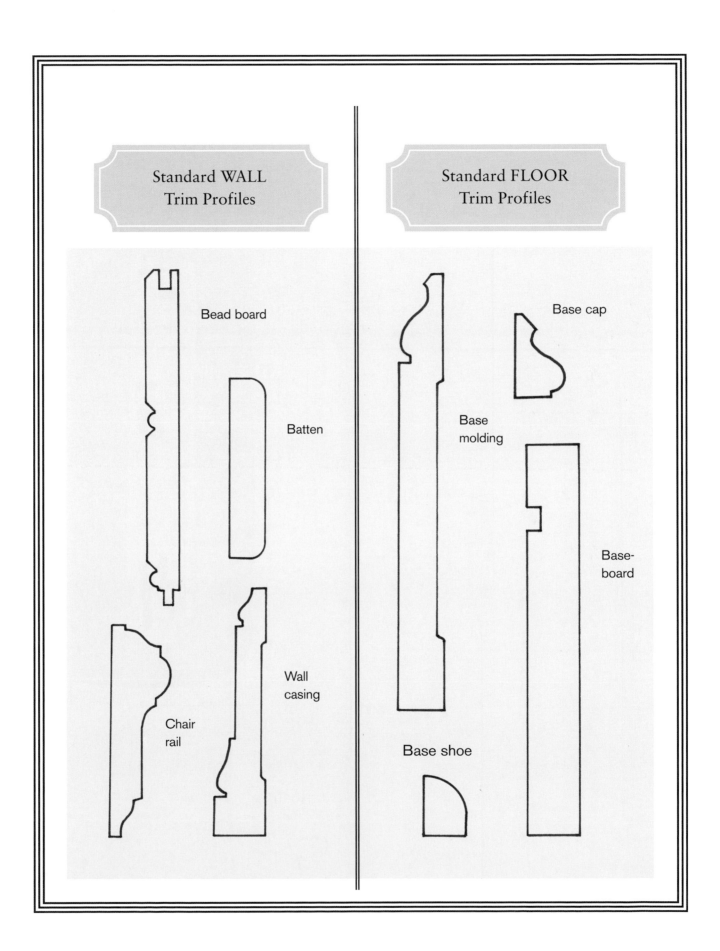

Standard WALL
Trim Profiles

Bead board

Batten

Chair
rail

Wall
casing

Standard FLOOR
Trim Profiles

Base cap

Base
molding

Base-
board

Base shoe

Arches often are used in formal
settings. New synthetic products on
the market make it easy to create the
arched effect.

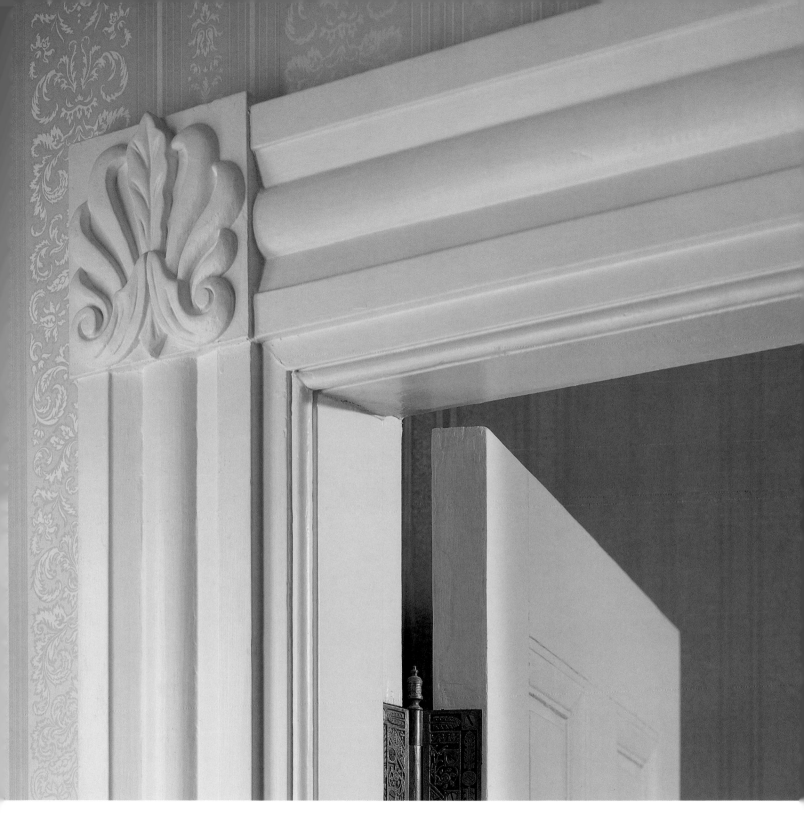

This Victorian-style door frame has a casing paired with a back band for a more decorative look as well as adding weight and thickness to the casing. Rosettes or corner blocks placed at the upper corners of the door also give an added touch of decoration.

Columns

As discussed earlier, columns form the basis for most trim profiles. As elegant additions to an interior, columns often are reserved for more formal living spaces. Columns fall into five classical orders: Doric, Ionic, Corinthian, Tuscan, and Composite (the first three are Greek and the last two Roman). The human body inspired the form of columns —starting with a solid base and slendering to a top or neck. In ancient times, as today, a building's purpose and status could be revealed through the type of columns it had. This was especially understood through its capital. The simple yet masculine Doric might front a handsome manse while an ornate, feminine Corinthian column might be used on a civic building or church. Although the proportion of columns was worked out over centuries, there is no set formula for any given order—only relationships between elements that takes study and experience to execute well. Today, various materials are used to make interior columns such as wood, glass-fiber-reinforced concrete, fiberglass, and plaster. Many architects use columns in conjunction with the existing molding profiles used. Although there are no rules to adding columns to a house, a few common mistakes are made when choosing or installing columns. Some of the biggest gaffes are not aligning the top of the column with the architrave; adding columns that are too narrow and out of proportion with the rest of the room; misusing a base for a capital or a capital for a base; no thought to classical proportion and context; and the overuse of the column in the house. Many architects agree that when you overuse columns throughout the home, no room is special.

Right: The Ionic column screen creates a visual divider between the dining room and the hallway of this open floor plan on the first floor of this beach house.

Left: The Corinthian column acts as a divider between a master bedroom and a sitting room.

Pilasters are based on the same
classical orders as columns.

Pilasters

As defined by architectural historian James Stevens
Curl, a column is an upright, circular element with a
base (except for Doric), a shaft, and a capital. It tapers
towards the top and supports the pediment. A pilaster
is shallow and rectangular in form and projects from a
wall. The most common places that pilasters are used
are around door casings and fireplaces.

Column Capitals

Capitals—the topmost structural member of a
column—determine the style of a column. The
capital can be as decorative as the Corinthian or
as simple as the Doric.

The design of the column capital is based some
what on the human form. The Ionic capital, with
its two scrolls, represents a popular hairstyle for
women in ancient Greece. The Doric capital and
column shaft reflects the strength of the male
form. The Corinthian column is the most ornate—
its capital has flourishes of acanthus leaves and
volutes. This grand style of column is most often
used in civic buildings. Not to be out done by the
Greeks, the Romans created their own set of
columns—the Composite and the Tuscan. The
Composite order combines both the Corinthian
and Ionic design, creating a more embellished
design. The Tuscan capital reflects a simplified
version of the Doric order. Although columns
come from the ancient world, they are used over
and over today in interior residential design as
support members or decorative room dividers.

Corinthian (Greek)

Columns are often misused or over-used in home design and should be reserved for special areas in the home. Columns come not only in wood but also in glass fiber-reinforced concrete.

Ionic (Greek)

Doric (Greek)

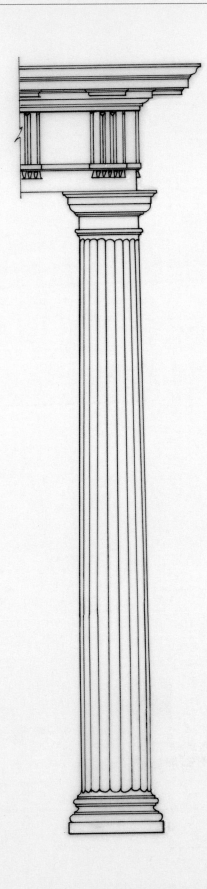

Composite (Roman)

Tuscan (Roman)

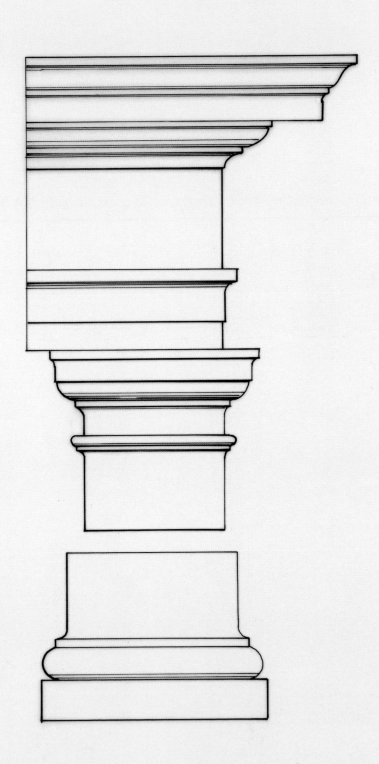

The built-up molding around the interior of this front door adds rich detailing based on historical forms.

Mantels

A frame for the fireplace, the mantel is an important part of the overall aesthetic in a room and is quite often the focal point to the design. The mantel is a symbol of hearth and home. Architectural historian James C. Massey points out that mantel trim often resembles an entryway's built-up door frame, such as the one shown left, because it is a portal of another sort—a frame for the flickering flames.

Decorative mantels became a common feature in home design in the 1700s and often set the focal point for a room.

Built-Ins

A built-in, or permanent furniture as it once was called, is an ingenious way to create more storage or seating space within the home. In vogue during the early 1900s, built-ins offered a streamlined, functional alternative to gaudy Victorian furnishings. They also were considered more hygienic than freestanding furniture because dirt and germs had nothing to hide underneath or behind since the piece was affixed to the house. Inglenooks, breakfast nooks, cabinetry, cubbies, sideboards, and benches all fall under the rubric of the built-in. Many architects carry on this tradition by incorporating built-ins into today's home designs. Although one of their main functions is to create additional storage and make the most of a space, a built-in can become an integral part of the overall design of a room.

The most common type of built-in is used for seating— typically placed against a wall, or a window, or even a bank of windows. Built-ins offer a perfect spot to curl up and read or just daydream. Benches also do double duty as storage, their seats hinged to offer a toy box or blanket chest. Breakfast nooks are also favorite built-ins. The table in a breakfast nook is quite often a fixed table that can be folded up to create more floor space while benches can be fixed or unfixed. The inglenook, introduced in medieval Europe, became a high-style fixture of Colonial Revival and Arts and Crafts houses of the early 1900s. A pair of built-in benches around a fireplace, the inglenook brought the hearth back to the center of attention in the home.

Another form of built-in and a feature we could not do without today is cabinetry. Built-in cabinetry came into fashion in Victorian-era butler's pantries and became a mainstay in kitchens by the 1920s. Colonnades, low cabinets that divide two spaces within a home (for example, an entryway and a living room) are also gaining popularity in home design once again. Although the cabinetry has remained a constant in home design, many types of built-ins are just coming back into fashion, as architects look to past forms for fresh ideas in home design.

Create Hierarchy

Designer Christine G. H. Franck often creates a hierarchy of trim within the home. She will break the types of trim into three families of design and call them A, B, and C—family, A being the most detailed and C being the simplest. The trim will have similar profiles but each family will relate to the space in which it is being used. For instance, living and dining rooms, which have 10' (3 m) ceilings, will be outfitted with family A, the most elaborate trim. The master bedroom and second floor hall receive family B—elaborate but more scaled to the room's ceiling height of 8' (2.4 m). Pantries, powder rooms, and guest bedrooms will be finished with family C, which is the simplest of the three groups, and typically made up of standard sizes you can pick up from your local home center.

A seating nook spaced in a semicircle is part of a stairway. The clever pairing combines two architectural features while saving space within the room.

Staircases

A staircase—a flight of stairs, which includes supports, handrails, and framework, and all their components—has many vertical and horizontal elements that fall under the rubric of trim. A staircase consisting of fine woodwork often can be the focal point or most stylistic feature to a home. The newel post, the solid central post that provides support of a staircase, is often the most decorative woodwork. The balusters, a number of vertical members used to support the handrail, can be as simple or as decorative as the design allows. Handrails themselves can have beaded edges to offer the decorative and practical purpose of making the smooth wood easier to grasp. Even the tread and risers can add style to the stair. Are they painted or left in a natural wood finish? Only a skilled carpenter should assemble a staircase and all its many precise parts. But all the same, they add visual detail along with practicality—getting from one floor to another in your home!

Stairway Anatomy

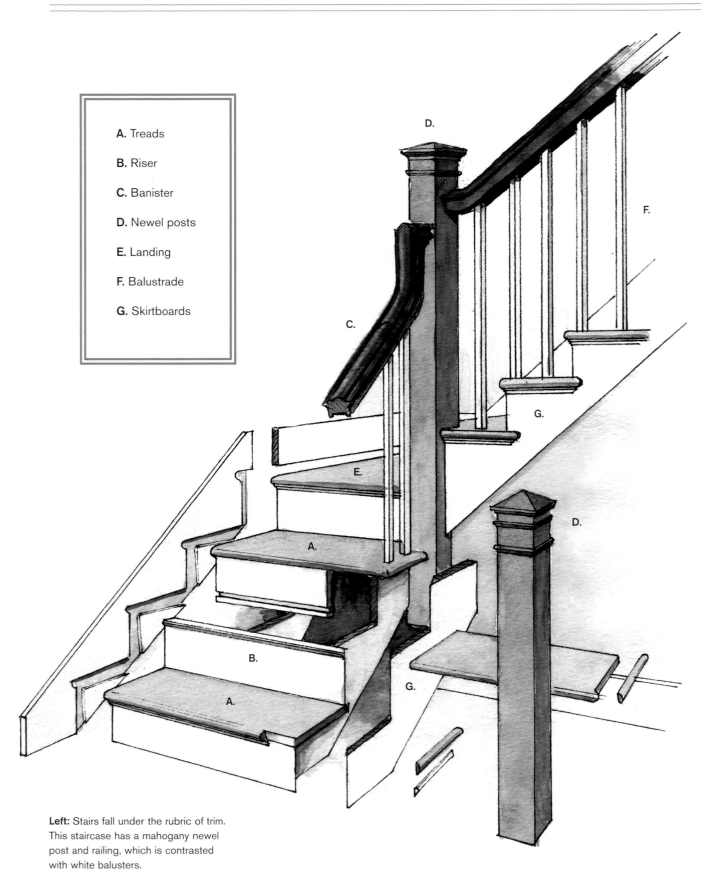

A. Treads

B. Riser

C. Banister

D. Newel posts

E. Landing

F. Balustrade

G. Skirtboards

Left: Stairs fall under the rubric of trim. This staircase has a mahogany newel post and railing, which is contrasted with white balusters.

Period Styles

Throughout the history of house design, trim has been an important component to defining the architectural style. We'll look at the history and hallmarks of six distinct period styles that are still being used in home design today. Whether you are adding trim to a neoclassical house or to an Arts and Crafts bungalow, there is an appropriate style of trim for your project.

Georgian: 1700 to 1780

Georgian architecture was introduced in the early 1700s in England. A revival of sixteenth-century Renaissance architect Andrea Palladio's classical forms, the style became a stronghold in both England and Colonial America. This style was brought over by English loyalists who were seeking a way to express their newfound wealth and status as landowners and merchants in the New World; they wanted to express their taste and knowledge of architecture in the buildings in which they lived. Architectural pattern books were the most important influence on the building practices at the time. James Gibbs' *Book of Architecture* (1728) had a great influence on American building design. Gibbs' book highlighted the features of classical design for both the interior and the exterior of the house. Carpenter's handbooks were particularly important in America because there were few trained architects living in the colonies. Carpenter-builders relied on these books, which did more than outline the exterior elevation. They were full of the interior details and the trim work that gave these

homes their Georgian style—door and window casings, mantelpieces, cornices, baseboards, columns, and chair rails. The most prolific author of carpenter handbooks was Batty Langley, whose books had a blend of geometry, classical orders, and good taste that appealed to builders.

At first, the classical order was timid and subdued. But as the design vocabulary stretched its legs, the designs became more elaborate. Elements of moldings in a Georgian room typically consist of the crown, chair rail, baseboard, and casings. Although exterior treatment was of strict Palladian rule, the interiors were freer in expression and offered details in Baroque and even Rococo ornament. Walls were often treated in decorative wood paneling. After 1750, paneling often was found only on the lower half of the wall—from the chair rail (also called the dado) to the floor. Important doors frequently received a frieze and cornice. Window casings were often eared at the top, much like doors. Ceiling cornices generally were made of wood and plaster and could range from simple cornices to elaborate entablatures with detailed dentils and modillions. Other details included motifs in birds, animals, and pastoral scenes. Georgian classical trim was made mainly out of wood because it was easy to carve into intricate moldings. An important new material was fine, white lime plaster. Ceiling beams were covered in this new white material, and it often was molded into cornices.

The Georgian style of moldings is based on classical architectural forms and was prominent in both English and American architecture in the eighteenth century.

Federal: 1780 to 1820

With the birth of a new Republican nation came the need for a new architectural style. The Federal style (1780 to 1820), as it is known, pays strict adherence to a revived Roman classicism. At first impression, the difference between the Georgian and Federal styles is not striking —both use the classical vocabulary in similar ways. But a second look at the style reveals a closer adherence to the classical form, both proportionally and through its details.

The influence came from sweeping social changes throughout the Western world. Architectural archeology discoveries in the 1700s sparked a renewed interest in Roman architecture. English architects Robert and James Adam, who had the largest architectural practice in England during the period, translated the classical style to a high art form. Graceful details and more vertical proportions came into fashion. The ratio of Federal style was 1:7, as opposed to the earlier Georgian ratio of 1:5. This ratio created delicate vertical forms within rooms. As in the Georgian style, moldings consisted of baseboard, chair rail, cornice, and decorative window and door casings. Ornaments often included swags, urns, garlands, eagles, and geometric shapes; patterns were mostly elliptical, circular, or fan-like shapes formed by fluted, radiating lines. The scale is smaller in Federal-style details compared to Georgian style, where the scale of the structural parts, such as in doors and windows is larger. The Federal style was a refinement of the earlier Georgian style established by the merchant wealthy. The most notable architects were Charles Bulfinch, Benjamin Latrobe, Samuel McIntire, and Alexander Parris.

During the Federal period, moldings became more vertical and delicate and followed the proportions of classical architecture more closely.

Georgian Style

If you have a Georgian-style home, select trim that incorporates its hallmarks:

CLASSICAL STYLING

THICK PROFILES

ORNAMENTAL DETAILING ABOVE MANTELS

BIRD AND ANIMAL MOTIFS

PASTORAL MOTIFS

WALL PANELING

Federal Style

If you have a Federal-style home, select trim that incorporates its hallmarks:

DELICATE DETAILS

VERTICAL PROPORTIONS

EAGLE DECORATION

SWAG MOTIF

URN MOTIF

PILASTERS AROUND DOORS AND FIREPLACES

Greek Revival: 1820 to 1860

The Greek Revival style was a predominant style of architecture between 1820 and 1860. Interests shifted from Roman architecture to Greek architecture during a time when the Greeks were at war with the Ottoman Empire. This heightened the belief that ancient Greece was the cradle of democracy and America embodied that democratic spirit. The Greek Revival house, in its fullest sense, looks nothing less than a Greek temple, with columns supporting an entablature. The form was realized in both farmhouses and grand houses alike. With the wide distribution of pattern books, the style spread out across the United States. Again, the carpenter's handbooks showed not only elevations but also the interior details. Two examples are Asher Benjamin's *The Architect or the Practical House Carpenter* (1830) and Minard Lafever's *In the Beauty of Modern Architecture* (1835). In addition to the carpenter's books, which featured Greek Revival building methods, several architects were well versed in the new style: Benjamin Latrobe, Alexander Parris, and Isaiah Rogers, to name a few. Greek Revival trim work has a commanding permanent presence. One of the most prominent details of the Greek Revival house is its columns. Greek Revival columns are of the Greek order: the Doric, Ionic, and Corinthian. The column is defined by the details found in its capital. The Doric is simple and strong; the Ionic column curls on it sides, which represent a woman's hairstyle; and the Corinthian sprouts acanthus leaves. Interior decorative details include the egg and dart, the Greek key, fret, honeysuckle leaf and cresting. These details can often be found on door and window crowns.

Greek Revival Style

If you have a Greek Revival-style home, select trim that incorporates its hallmarks:

CLASSICAL GREEK DETAILING

GREEK KEY MOTIF

FRET MOTIF

EGG-AND-DART MOTIF

USE OF COLUMNS

A common characteristic of the Greek Revival molding style is the dog-eared doorframe—where the top comes about almost in a T-shape.

Victorian: 1860 to 1900

The Victorian era ushered in the advent of the industrial revolution when moldings and trim went from being hand carved to machine milled. Other sweeping changes came in the way of the style of homes and their trim. During the Victorian era, there was a departure from the stoic proportions of classical architecture to a freer form in design. Moldings became even more elaborate during the Victorian era, correlating to the era's housing stock, including Queen Anne, Italianate, and Eastlake. These new, fanciful trim designs could be built of up to 20 components! More was definitely better during this era. Most of the trim being produced at this time came from factories. By 1890, millwork catalogs offered all components of woodwork for the home and comprised all kinds of interior elements, such as doors, windows, transom frames, stair parts, and moldings. Wood was plentiful, and inexpensive pieces were elaborately carved by machine with decorative motifs such as Japanesque and floral motifs. Because room sizes were often bigger and the ceiling height taller during this time, base moldings were taller—often 9" to 12" (22.9 to 30.5 cm). Victorian moldings were heavily ornamented and often had a combination of elements such as rosettes and header blocks. By 1914, Victorian moldings had all but disappeared from millwork catalogs, as a new era of architectural style was brought to the housing market—the Arts and Crafts house.

Victorian Style

If you have a Victorian-style home, select trim that incorporates its hallmarks:

ELABORATE DETAILING

MACHINED MILLWORK

JAPANESQUE AND FLORAL MOTIFS

ROSETTES

HEADER BLOCKS

Victorian door casings were often punctuated with rosettes at the top corners of the doors. During the Victorian era, millwork was mass-produced and came in a wide variety of shapes, sizes, and decorative embellishments.

Arts and Crafts/ Bungalow: 1900 to 1930

By 1900, the Arts and Crafts movement had reached across the sea from England to the United States. The movement started by designer William Morris rejected all the gaudy and ostentatious affects of the Victorian era—rooms were scaled down and so were the moldings that went in them. Sizes of moldings became shorter, from 4" to 6" (10.2 to 15.2 cm), which was in keeping with the smaller proportions of the Arts and Crafts room.

Although less ornate than the built-up Victorian moldings, Arts and Crafts trim had a profound effect on the overall room design. Molding profiles and millwork were simple, and unadorned, and often finished in a dark stain to enhance the beauty of the wood, typically quartersawn oak, which was chosen for its prominent grain patterns. Other wood species used in Arts and Crafts interiors included fir, red gum, red pine, and cypress. This practice of bare wood played up the tenet of the Arts and Crafts philosophy: "to have nothing in your house that is not believed to be useful or beautiful." Handcraft or the appearance of handcraft also was an important component to Arts and Crafts design.

The woodwork within the Arts and Crafts house, as in other period styles, unified the spaces as a whole. Walls were often paneled in high wainscoting, and plate and picture rails were common. Built-ins were abundant, particularly colonnades, inglenooks, and china cabinets. The wood trim was carried throughout the house for continuity. For example, the profile of the crown molding would also appear in the mantel. Arts and Crafts trim exudes warm tones, a sense of masculinity, and chunky proportions, which all give the trim work its distinct look. The organic nature of the trim design gives it a visual connection to the natural world. As with many of our period styles, the trim work within the house was a defining feature of the Arts and Crafts architecture. Trim work often mimicked the furniture styles of the day, and scale and proportion were again critical elements.

Arts and Crafts Style

If you have an Arts-and-Crafts-style home, select trim that incorporates its hallmarks:

SIMPLE DETAILING

UNADORNED TRIM

NATURAL WOOD TONES

HIGH WAINSCOTING

PICTURE RAILS

PLATE RAILS

The Arts and Crafts design was simple, often with flat, unadorned profiles, which was a departure from the Victorian aesthetic. Plate rails were a decorative and practical embellishment to the home.

Revival Styles: 1900 to 1940

As the Arts and Crafts era discarded classical motifs and ornamentation, Revival styles of the early- to mid-twentieth century turned back to the colonial era and more classical forms popped up in the design of interiors, particularly the home's moldings. The most predominant revival style was the Colonial style. Most Colonial Revival trim is almost identical to the earlier Georgian or Adam styles of the 1700s. These early styles formed the backbone of the style. Postmedieval English or Dutch also were influences on the Revival style. Unlike its earlier prototypes, Revival trim and moldings were machine milled—the handcraft was no longer visible in the trim work. Box-beam ceilings, window seats, and arched door frames were all hallmarks of the Revival-style home. Formal rooms such as studies or living room walls, often were covered floor to ceiling with traditional raised paneling. Colonial Revival ceiling heights once again allowed for more elaborate moldings. Plaster frequently was used for crowns or even fireplace surrounds. Today, some contractors use plaster for decking out rooms with trim. Plaster can be shaped easily, it dries quickly, and it is often easier to install than wood molding.

Colonial Revival Style

If you have a Revival-style home, select trim that incorporates its hallmarks:

CLASSICAL DETAILING

PLASTER AND WOOD

BOX-BEAM CEILINGS

ARCHES

WAINSCOTING

Popular at the turn of the twentieth century, the Colonial Revival style has made a strong comeback in home design and is often referred to as the Neoclassical style by today's home builders. The style is based on forms adopted by Colonial America from ancient Greece and Rome.

Don't Mix and Match

One rule of thumb when adding trim to your home is to keep the profiles consistent throughout the rooms of the house. This will create continuity in the overall design. Be careful not to mix and match opposing styles. For example, you wouldn't want to pair a classical crown molding with Arts and Crafts wainscoting. Just as these styles wouldn't work well together, neither would high-style Victorian moldings meld in a Colonial Revival house. Before choosing the style of moldings for your rooms, look at the overall design of your home. Another factor to consider is the hierarchy of rooms within your home. Service rooms such as the kitchen, laundry room, or the pantry typically receive simpler profiles while living spaces such as entry halls, great rooms, dining rooms, and master bedrooms receive more elaborate profiles.

Your trim finish doesn't have to be painted stark white. Today, designers are branching out and using any number of paint colors and stains for the perfect finishing touch. When you are painting trim, one design principle to keep in mind is to create continuity in the design.

CHAPTER Three

Materials and Finishes

We've explored trim types, shapes, and period style, but what about the kinds of materials and finishes used for the trim in your home? Wood is still the most popular choice when it comes to dressing walls and ceilings with trim, and there are stock and standard options at your home center or lumberyard. When you are working with a custom mill shop, your choices of shape, wood species, and wood grade are vast.

Cast plaster is also an alternative choice with builders. Because of its viscous nature, plaster is easy to form into any ornamental shape and cures quickly but it should be applied by a professional.

There are also synthetic materials for trim, such as medium-density fiberboard, that are good alternatives to wood. Primed for painting, once installed and painted, fiberboard looks like real wood trim. It is a composite of wood fibers and resin, is smooth, and often requires little sanding.

Polyurethane trim is an inexpensive alternative to wood trim. It is lightweight and, since it is made from molds, comes in a variety of patterns. This wonder material does not require priming or sanding before painting and it won't crack, split, splinter, or rot.

Flexible vinyl is just that—a material that can bend to any shape you need it to, making it great for archways and tight curves. The vinyl is cut and installed just like wood trim but is not affected by moisture the way wood is.

This arch is actually a synthetic material that makes it easy to flex and is a less expensive alternative to a custom-kerfed arch. The other trim elements in the space are poplar.

Wood Types for Trim

Wood for architectural interior details, including trim, comes in dozens of wood species. Also in the wood trim category is veneered plywood. Along with solid wood, there is wood chip, particle, and wood fiber bonded with adhesives. The wood is graded based on its appearance, and there are two general categories of wood: softwoods (coniferous with needles) and hardwoods (deciduous and leaf bearing). Generally, hardwoods are harder than softwoods but this is not always true. Softwoods generally are used in lightwood framing while hardwoods often are used for finish carpentry. Exceptions to this rule are pine, cedar, and redwood, which are all softwoods and are routinely used in finish-grade millwork. Along with durability for a particular use, other factors to consider are a wood species' availability, your personal taste, house style, local traditions, and, of course, cost.

Another factor when purchasing your wood trim is how the wood was cut. When logs are cut, they are often sawn with a use in mind to make the most of the yield. The orientation of the wood cut relative to the direction of the growth rings determines a variety of lumber characteristics, such as stability and grain direction. Since wood is likely to shrink along a tree's growth rings, a plain-sawn board (tangentially cut) will shrink more and likely warp more than a quartersawn board (radically cut) from the same log. Because of this, quartersawn hardwood and vertical-grained softwood are highly valued for their stability. Knots, splits, and a short grain can determine the grade. The grade or appearance is important when the wood is left natural or receiving a stain only where the wood will show through. When choosing your wood for your trim project, look at the species, which will often determine the aesthetic effect.

Common Woods Used in Trim Today

Softwoods

DOUGLAS FIR

PINE SOUTHERN YELLOW

REDWOOD, HEART

SPRUCE, WHITE

Hardwoods

CHERRY

WHITE BIRCH

YELLOW MAPLE

MAHOGANY

POPLAR

OAK

RED OAK

WHITE OAK

Hardwoods, such as the rich cherry used on this wainscoting and overmantel, are often used in finish carpentry.

Custom Millwork

Going to your local lumberyard or home center to pick up stock trim is an easy task. Ordering custom hardwood is another story altogether. Custom hardwood comes in rough rather than standard, even lengths and widths. It is also unjointed and unplaned. In addition, depending upon what grade the board is, it may need knots or wane taken out. With all these factors to consider, working with a contractor who knows his or her way around a custom mill shop is best. You might even want to take your contractor to a local lumberyard. To get the most from your rough lumber, make a list for the mill shop as to what you are looking for, such as quantity, width, length, thickness, finished edges, finished faces, sanded faces, and species. Also note what is acceptable in terms of sapwood, gum, and moisture content.

Trim Talk

- Sapwood is the wood between the bark and the heartwood—it is often lighter in color than heartwood.

- Gum is a gooey wood substance that can leave dark blotches in the wood.

Check how many boards will need to be color matched and specify the size of each individual piece—and they should be cut to this size.

Wood Strength

The Architectural Woodwork Institute differentiates wood trim according to its length. Standing wood trim is trim that can be accommodated easily with single lengths of wood depending on species, such as crown moldings. Running trim is made up of finger-jointed wood to achieve the longer lengths customarily needed for this type of trim.

If your contractor or architect has specified custom millwork, knives will have to be made to create the desired profiles. There are mill shops that will do custom-trim profiles, and, oftentimes, depending on how elaborate the trim is, the cost will not be much more than stock trim.

Hardwood Characteristics

WOOD	DESCRIPTION
CHERRY	There can be great variance in color and grain between the heartwood and sapwood in a cherry plank. Cherry also contains black streaks or gum. If you specify minimal gum, this increases the cost significantly.
RED OAK	Red oak ranges from creamy white to pale brown in color.
HARD MAPLE	Hard maple's sapwood is light colored and because of this is a preferred choice for woodwork.
ASH	Similar in grain pattern to oak, its sapwood is light, which makes it a good choice for blonde woodwork.
BIRCH	Birch is hard, dense, and similar to maple. The heartwood is preferred.

When you are choosing a wood type for your trim, consider several factors. Will the grain be showing through a clear stain or will it be painted? Is the trim meant to be fancy or simple? What about the cost of the wood to be used? What is locally available? We'll look at some of the wood types used for trim in the home.

PRO	CON
Adds warmth and richness to a space.	Cherry can be very costly; it also darkens quickly when exposed to sunlight.
Popular wood is often used in Arts and Crafts houses. Takes stain well.	If you request minimal mineral staining, the cost for the trim will be higher.
Takes stain well and is heavy, hard, and strong.	Can be expensive.
Great for contemporary design. Excellent blending qualities.	Because of large pores, ash cannot be painted.
Its light color makes it ideal for modern room designs.	It is not readily available across the country.

Finishes for Trim

When you have chosen a good-quality wood, protect its surface with a good-quality finish. A finish will also enhance your trim work's color and warmth. One of the most important reasons for using a finish on trim is to seal out moisture, which can be caused by seasonal changes in the air. Moisture can warp or even split the wood's surface. By sealing the surface of the wood against the degrading effects of moisture, you will also help your finish last longer!

Several woods, such as Western red cedar and teak, have natural oils that protect them but most wood species do not, so coat your trim with a wood finish once it is installed. There are several finishes on the market but read the labels carefully. Some, such as stain and varnish, are just for reviving old finishes while others are for use on bare wood. Traditional treatments for finishing trim work have been proven effective and are still in use today, although modern technology has introduced a number of new potions to protect and enhance your trim work.

Not only do finishes come in a variety of formulas—stains, shellac, lacquer—but they also come in a variety of final lusters—flat, eggshell, satin, semigloss, high gloss, and every varying degree in between. But remember, it is not just the sheen but the type of finish used that will create a certain effect. For instance, a varnish and polyurethane will look very different even though they may both offer a satin finish. Durability should also be a factor when choosing your finish. Will the stain go on a crown molding where no one will touch it, or is it for a built-in cabinet, that will get a lot of wear and tear?

Before finishing any trim work, make sure it has been planed and sanded to create a smooth surface. The final sanding should be done in the direction of the grain—never against it. This will prevent splinters. Stain can actually call out imperfections such as scratches across the grain, rather than hide them. Use a 220-grit sandpaper and use a dust brush to clean any surfaces that will receive any water-based products. (If you are painting your trim, the surface doesn't have to be perfectly smooth.)

Trim Finish Options

STAIN offers richness to your wood's appearance. Shellac is a clear-coat finish.

LACQUER, although beautiful, is hard to apply and should be left to a professional.

POLYURETHANE is a clear-coat finish and is the most popular clear trim finish today.

PAINT covers the wood grain completely. The most popular paint colors for trim are white and cream.

This living room with pine trim was finished in a clear coat of water-based polyurethane to protect it from the coastal environment.

Stain

Enhancing the beauty of natural wood by staining is a popular way to treat your trim work and is the finish we associate the most with achieving that natural wood luster. The grain of wood can be enhanced by the use of stain. The stain won't hide the grain like paint but will enhance its color. Stain's primary function is often to create a deeper, darker appearance to your wood. Wood is a variable material, meaning that it changes and has irregularities that must be taken into account when finishing it. For instance, a wood conditioner might be needed to treat the surface so the stain will go on evenly and not blotchy. Some woods such as mahogany take stain much more readily. Stain is applied with brushes, by hand with a rag, rollers, and spray systems. Once you apply a stain, you should wipe its surface with a dust-free cloth to take off any excess stain. Before adding a clear finish, let the stain dry thoroughly.

Trade Tip

One note of caution: penetrating-oil stains are powders dissolved in oil that penetrate wood more quickly and dry a lot more quickly than water-based products. These include gasoline, Benzine, and naptha. All are volatile solvents that should be disposed of appropriately. When discarding cloths soaked in an oil-penetrating stain, soak in a water bath.

Shellac

Shellac is a clear-coat evaporative finish, which means it is a mixture of solvent and resin. The resin for shellac comes from the excretion of the shellac beetle. The resin is dissolved in alcohol, which leaves the hardened shellac behind as the alcohol evaporates. A fast-drying finish, it is compatible with a number of other finishes and makes an ideal base coat. Shellac is more difficult to apply than an oil-based finish but offers more protection. It offers a warm color and has been used since medieval times to protect furniture. In the more recent past, it was often used as a finish on Victorian trim work. Shellac comes in a few shades: orange for darker woods; white or bleached shellac, which yields a pale color; or dewaxed shellac, which is clear. When you are buying shellac, buy it in a thicker solution and thin it as you need to. Shellac is more fluid than varnish and will dry more quickly. Shellac should be brushed on in long, even strokes. Don't go over an area more than necessary; this will create a buildup in the shellac. Shellac should not be used in high-moisture areas because water can spot the finish.

Lacquer

Traditional lacquering is an arduous process but today's lacquers are mostly synthetic solutions that are clear or colored and dry quickly—and are a lot easier to work with. Lacquer comes from cotton fibers treated with nitric and sulfuric acids. The solvents for making lacquer are again highly volatile and should be disposed of with great care. When applying lacquer, you should use a large brush and stroke it on rapidly in one direction, even though it is often better to spray lacquer on. Because of the substance's volatility, a professional painter should apply it. If a synthetic lacquer needs to be thinned, use only the thinner that is supplied for that particular lacquer. The main problems with lacquer are its flammability and toxic fumes.

Varnish and Polyurethane

These are the most common clear finishes used today by both homeowners and professionals alike. The biggest issue with working with these finishes is their slow drying time. Because of the slow drying time, dust can become a factor. These finishes will also yellow over time, creating an undesirable effect. A combination of vegetable oils and resins with alcohol and acid are the base for most varnishes today. These alkyd resins have a high resistance to heat, water, and chemicals and are easy to brush on. Polyurethane products, available to both consumers and professionals, are a one-part product that contains an alkyd varnish modified with polyurethane resin. These one-part polyurethanes have the same good characteristics that varnishes have and are applied the same way but again, dust is an issue because they are slow drying.

Gaining popularity in the home finishes market are water-based varnishes. With the number of environmental concerns, the coatings industry is looking at new, more environmentally friendly finishes. They are created using the same technology as latex paints, should be applied heavily, and should never be thinned. Some of these water-based finishes are scuff resistant, quick drying, and don't give off a strong odor. They are also resistant to heat, water, and chemicals. When you are applying water-based finishes, they need to go on quickly and evenly to avoid any brush marks. These products also clean up with water so they make the job a breeze for the person doing the work. Never apply polyurethane over shellac; it will not adhere properly.

Applying the Finish

All clear finishes are applied by brushing, spraying, or wiping on with a cloth. For trim such as door and window casings and baseboards, a minimum of four coats of finish should be applied. Professionals also recommend three coats for wall paneling and ceiling beams. Bristle brushes are used when it is a solvent-based product while a nylon brush is ideal for water-based finishes. If you are tackling a large job, spray on the finish; this is best left to a professional. After each application is dry, sand with 220-grit sandpaper. Never use steel wool with water-based finishes. Then dust, vacuum, and wipe the surface with a rag before applying the next coat.

Trade Tip

Don't fill all the nail holes before applying the first coat of finish; you might get the filler in the surrounding wood and create a blemish. Apply the first coat of finish and then fill the holes. This will create a less noticeable patch. Use colored putty for a flawless effect.

Trade Tip

The best finishes for trim work are oil-based polyurethane varnish, water-based polyurethane varnish, conventional varnish, and lacquer.

Paint

Of all the finishes that can be applied to your trim work, only paint will cover the grain totally. Paint comes in latex, water-based, and oil-based types. There are pros and cons to both paints, but new environmental regulations are making it tough on oil-based paints. The pros for latex paints are that they clean up with soap and water, dry quickly, and are safer for the environment. But there are different grades of latex paints. The best have 100 percent acrylic resin; latex paints containing only vinyl resins are of lesser quality. Oil-based paints offer a smoother finish but can be harmful to the environment if not disposed of properly.

Alkyd paints dry without brush marks and are a great choice when finishing trim. Because of the quality of these high-gloss paints, they are often used to finish trim work. When you are using alkyd paints, wear a respirator.

Paint is typically used on inferior wood types. Before you apply paint, woods need to be properly prepped.

Softwoods, such as pine, need special attention because they can have knots on the surface that are resinous; these knots often can ooze resin, which will spoil a good paint job. To ensure that this won't happen, knots are often prepped with shellac. The success of any good paint job relies on the prep work and what lies underneath the paint. The wood should first be painted with a primer. You can also purchase wood that has already been primed. The primer actually bonds with the wood by penetrating the surface. This first layer provides a base to build upon with subsequent coats. The next layer is the undercoat and should be the same color as the topcoat—or at least a close match. The quality of your undercoat determines the quality of the topcoat. So although you don't need to be as careful with your paint job as you are with varnishing, you do want to watch for dust and debris getting into the paint. Normally, you will need only one topcoat. Apply this topcoat with a minimum of brush strokes to get a good even finish.

Trade Tip

To get that glossy, rich, oil-like quality from latex paint, use the proper brush. Use a high-quality, straight, nylon brush and change your brush often.

On this adobe, the designer chose a brilliant blue for the arched door casing. As long as the color doesn't clash with the design, your paint choices can be endless when it comes to finishing your trim work.

Painting Trim

Semigloss or gloss is the most common type of paint used for finishing trim. You can paint your trim before you install it and then touch it up; or you can install it and then paint. If you are painting the trim after it is installed, cover adjacent surfaces with masking tape. Paint your trim starting from the ceiling down. The width of your trim will determine what type of paintbrush you use. For example, for narrow trim use, a 1½" (3.8 cm) sash brush; for wider trim, use a 2" (5.1 cm) trim brush.

Staining Woodwork

To enhance the characteristics of the wood, many homeowners and builders choose to finish trim with clear finishes or stains, as discussed earlier. Softwoods absorb stain much more quickly than hardwoods. Softwoods also take stain more unevenly than hardwoods, so it is best to prep bare wood with a wood conditioner. To apply stain, wear latex gloves and use a brush or a clean T-shirt to apply the stain. Always follow the manufacturer's instructions to get the best quality job.

Trade Tip

PRIME: Sand wood smooth, apply a wood conditioner, and let dry for 24 hours.

FINISH COAT: Apply one coat of stain, either water- or oil-based, according to the manufacturer's directions and let dry. Sand the surface and apply another coat. Continue the process a third time if needed.

Apply a clear finish after the stain is dry.

Trade Tip

If you are trying to paint with latex over an existing oil paint, it might have a hard time adhering. Clean the surface of the oil paint, sand the surface with 15-grit sandpaper, wipe down with a clean damp cloth, and prime with a fast-drying latex primer.

The crown molding and window casings in this Colonial-style room are finished in an olive green, appropriate to the style and age of the house.

Your trim doesn't have to be white. Break out and choose colors that are appropriate to your house style. The homeowner here chose a muted blue to paint the trim and built-ins in this Colonial-style house.

CHAPTER Four

How to Install Trim

Adding basic trim can transform a humdrum room into an elegant, beautiful space. Whether it's stately crown molding or casual bead board wainscoting, these details reflect your own personal style and add richness to the space. Although not every finish or carpentry job is for the do-it-yourselfer, the novice carpenter can take on a few jobs. This chapter covers the basic carpentry techniques and tools needed to create simple trim details that will enhance the design of your interiors. Power tools needed for most of these jobs can be expensive to purchase; for these smaller jobs, you might want to consider renting the equipment that you'll need to get the job done. If you're not comfortable cutting your own lumber, say for a bead board installation project, your local lumberyard or home center can do the job for you. Measure carefully before going—it's a hassle to run out of boards halfway through a project!

Before beginning your project, make a rough sketch of the room and determine where you are going to start to install the trim. Start with the moldings that you will see first upon entering a room and work from there. If you're lucky enough to work in a perfectly square or rectangular room, your trim installation will be fairly straightforward. But if you have an irregularly shaped room, for example, with a bay window, you should start the trim around that projection or recess. These rooms will involve more inside and outside corners, so it is wise to get a handle on the different joints these corners take. (See sidebar on joints, page 77.)

Trade Tip

It's important to attach heavy moldings such as plate rails to wall studs. But how do you find them? Wall studs are typically placed every 16" (40.6 cm) and a good place to start looking for them is around wall outlets; they are almost always placed there. You can also use a stud finder although many contractors simply tap on the wall and listen for a solid rather than a hollow sound. Moldings should be nailed into the studs whenever possible.

Bead board is a popular trim treatment in bathrooms and is fairly easy to install.

Trade Tip

When installing trim:

1. DON'T USE NUMBERS.

It is more accurate to hold a board in place to mark its length, rather than use a ruler.

2. AVOID A FLUSH EDGE.

Because wood moves, swells, and contracts, it is almost impossible to get a flush edge, for example, in a door casing.

3. STEP TRIM BACK.

Stepping trim back to show reveals will help create shadows and it will make it harder to see unevenness in your trim work.

4. COVER THE END GRAIN.

Paint and stain look different on the end grain of wood, so it is best to cut a mitered return to cover your end piece.

5. YOUR WORK DOESN'T HAVE TO BE PERFECT.

Put your time and energy into what will show in the end. For instance, if your wallboard doesn't come all the way to the floor, don't worry because your baseboard will cover it!

Safety Gear

The following is a list of materials you might want to consider picking up at your local hardware store:

EAR PROTECTION

SAFETY GLASSES

RESPIRATOR

PARTICLE MASK

WORK GLOVES

KNEEPADS

Types of Joints

Depending on where or how you are installing your trim work, the following text describes the different types of joints that you will need to know about when installing trim.

Butt Joint

A butt joint is a joint where two flat pieces of trim are positioned next to each other. The best place to use a butt joint is the place where two square profile pieces come together in an inside corner. The butt joint is also a good solution for simple window and door casings. For an attractive finish, get a tight fit between the two pieces of wood.

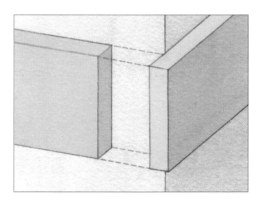

Mitered Joint

A mitered joint is a joint where two pieces meet at an angle. Think of a joint in a picture frame. A mitered joint is more desirable around window and door casings and offers a nice, tight fit.

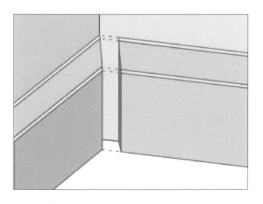

Coped Joint

When one piece of trim is butted into a corner and another piece is cut to fit against that trim profile, it is called a coped joint. The best places to use a coped joint are in the inside corners of chair rails, crown moldings, and baseboards. When the moldings are highly detailed, this is the best and really only way to achieve a tight joint.

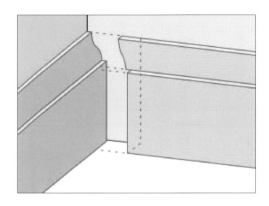

Scarfed Joint

When you are placing two moldings together on a long wall, use a scarf joint. If you butt the pieces, it will be noticeable. Cut the joining pieces at opposing angles. A 30-degree angle is best. Make sure the joint is angled away from the direction from which it will be viewed. This will create an almost seamless appearance in the trim.

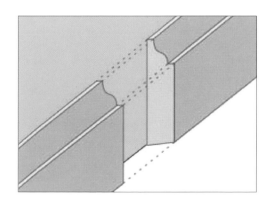

How to Install a Chair Rail

While adding a decorative touch, chair rails serve a very practical purpose—to protect your painted or papered walls from chairs bumping up against them. One question often asked is how high a chair rail should be. Good common sense says to measure the height of your chairs. Where does the top sit against the wall? Another safe bet is to place the chair rail anywhere between 32" and 36" (81.3 and 91.4 cm) from the floor—this is standard for 8' to 9' (2.4 to 2.7m) -ceilings and is about one third of a room's height. Although chair rail trim is available today at home centers, this commercial offering can look thin and anemic when installed. One good option is to build up your own profile by using select pieces. A good wood option for this type of trim is poplar or select pine. Add a cap and a cove molding, and you will have an ornate, wall-saving detail in your room.

Chair rails typically are placed about 3' (0.9 m) above the floor and protect the wall from the backs of chairs. This paneled wainscoting is topped with a chair rail.

How to Install a Chair Rail

Tools

- Level
- Chalk-line box and pencil
- Moldings
- Saw
- Hammer and nails or finish nailer
- Wood glue and wood filler
- Sandpaper

1.

Establish the height of the chair rail, and then use a level or chalk line to mark an even line across the wall where it will be installed.

2.

Once you have a line to work from, attach a board (such as a 1" x 4" [2.5 x 10.2 cm] cleat planed to 0.5" [1.3 cm]) below your line.

3.

Run a bead of glue along the cleat, which serves as the main rail to strengthen the connection to the cap piece.

4.

Measure and cut the cap to fit the wall and then fasten the cap to the rail by using a hammer and nails or a finish nailer.

5.

Measure and cut the cove molding to fit the wall.

6.

Finish the installation by attaching a cove molding below the cap. Fill the nail holes, sand surfaces, prime, and paint.

How to Install Bead Board

A popular building material in the Victorian era, bead board fell out of fashion in the mid-twentieth century—much the same way most trim details did in the advent of production houses. Bead board has made a strong come-back in the last few decades—first in coastal beach cottages and today in many traditionally styled kitchens and baths. The most common bead board is joined together by the boards' tongue and grooves to create one continuous, integrated wall or ceiling surface. Beads are milled right into the boards that help hide the joint. Common widths of bead board are 3" to 4" (7.6 to 10.2 cm) with 2¼" to 3¼" (5.7 to 7.7 cm) showing. Today, you can purchase sheets that mimic the look of bead board but don't offer the richness and quality you will find in single boards. The typical height of bead board is 36" (91.4 cm) although it can run higher on the wall—5' (1.5 m)—to create a lighter effect in rooms with high ceilings.

Trade Tip

Once you are several feet from a doorway, measure how far the casing is from the top and the bottom of the bead board. If there's a difference between these measurements, fan out successive pieces until they become parallel with the casing. Fit the last two full-width boards and measure the remaining distance to the casing. Cut the third piece to that width on a table saw. Put all three pieces together, bending them to form a slight curve, and press them between the casing and wainscot.

Adding a Flat Board Rail

An easy way to finish off the bead board is by adding a flat board to the top of the boards—much like baseboard. Start by fitting the rail between the inside corners. Miter and glue all joints. Then continue nailing the board around the room. This is a clean, simple solution to finishing off the bead board.

Bead board has been a popular wall treatment in bathrooms for more than a century. Today, there are bead board sheets that are easier to install than individual boards.

Step-by-Step Bead Board

Tools

- Cordless drill (and bits)
- Levels (2' and 4' [0.6 and 1.2 m])
- Compass (for scribing)
- Pencil
- Coping saw
- Biscuit joiner
- Utility knife
- Nail set
- Caulking gun
- Jigsaw
- Combination square (6" [15.2 cm])
- Tape measure
- Block plane
- Brad nailer or finish nailer
- Hammer (16 oz. [0.45 kg])

If you are cutting your own boards, a table saw and miter saw are needed.

1.
Using a 4' (1.2 m) level, place bead board vertically against the wall and pencil a layout line on the wall.

2.
Place one board on the wall and place another board alongside. Tap the boards together using a wood block.

3.
Secure the boards to the wall toenailing 4d nails through both tongues—top and bottom. Slip each board's groove over the tongue. (For extra security, add an adhesive to the wall, press wood into adhesive, and toenail the tongue, top and bottom. If you are hammering, drill pilot holes first and use a nail set.)

4.
Hold the same board alongside the outlet so that the board's top edge touches the layout line. Mark the side of the board where it touches the top and bottom corners of the outlet. With a combination square, draw horizontal lines out from the marks on the side and connect them to a vertical line running up from the bottom mark. The connected lines outline the notch where the outlet will go.

5.
Cut out the notch with a jigsaw and then nail the board as in step 3.

6.
To fit the boards when you are working in a corner, simply cut the bead off the board, and fit together.

Trim Talk

There are generally two ways to finish corners—either a mitered or a coped joint. Exterior joints are often mitered while interior joints are coped.

MITERED

A miter joint is a joint between two pieces of trim at an angle to each other. Each piece is cut at an angle equal to half the angle of the joint—the trim is typically at right angles to each other.

COPED

Fit the first piece on one wall with a square end butting into the corner. The end of the trim on the other wall is cut to fit against the shape of the profile of the molding on the first wall.

Trade Tip

Coping a piece of molding offers the best finished joint between moldings in interior corners, and requires skill. First, cut the end to be coped at a 45 degree angle, with the furthest point corresponding to the overall length of the trim piece. Then taking a cope saw, remove the exposed material by back beveling along the edge of the trim profile. This beveling creates a hollow pocket that accommodates the abutting trim. What remains is a hairline edge that matches perfectly to the profile of the already installed trim.

How to Install Crown Molding

Installing crown molding can add elegance and interest to a room. Despite the long runs of material, it is one of the more difficult projects to tackle, and is probably best left to the professional. Choosing the crown detail is subjective and can offer an interesting pursuit. A standard height for crown for 8' (2.4 m) ceilings is approximately 4" (11.4 cm). Room dimensions will dictate any further build up—a room with higher ceilings might require a more substantial trim profile, sometimes with multiple pieces in its overall makeup. When beginning your project, remember that the last piece of trim installed must fit tightly against the first piece installed. The first piece would require both ends to be square cut. Then working your way from right to left around the room, each successive piece is installed with the left end square cut and placed into the corner while the right end receives what is called a coped cut (see tip below). The last piece of crown molding installed is cut a bit longer, with both ends being coped. The two ends are installed first with the center bowed out into the room. Once the two ends are perfectly aligned with the adjoining pieces, the bow is pushed towards the wall, compressing the two coped ends tightly into the corners, leaving a very nice finish.

Crown molding is the trim that sits at the top of the wall. The crown molding shown here has dentil detailing.

How to Install Crown Molding

Tools

- Chalk line
- Pencil
- Standard 4" (10.2 cm) crown molding
- Nails

1.

Hold a scrap piece of crown molding at the correct angle on the wall and mark the spot. Hold the same scrap piece at the other end of the wall and make a mark.

2.

Next, measure the distance from the ceiling to the marks.

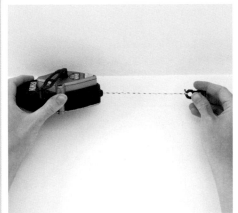

3.

Snap a chalk line between the marks.

4.

Apply the crown molding trim to the first wall with square ends in both corners.

5.

Use nails of sufficient length to penetrate into solid wood at least 1" (2.5 cm). Fasten the molding with a nail at both upper and lower edges. Nail at 16" (40.6 cm) intervals. End nails should be placed 3" (7.6 cm) from the end to prevent splitting.

6.

Cope the starting end of the first piece on each succeeding wall against the face of the last wall. Compress the two pieces into place. Add caulk or filler to any small gaps before finishing.

How to Install Baseboard

Baseboard is an important and practical design element in a room. Not only does it offer visual interest to a space and punctuate where the wall begins and the floor ends, but it also hides the seam between the floor and the wall. The baseboard helps protect the bottom of the wall from scuffmarks, and bumps, and dings from brooms and mops. It is a fairly easy trim component to install but you will want to invest in a pair of kneepads for this job!

The height of your ceiling and the dimensions of your room will determine how thick your baseboard will need to be. For instance, if you have 10-foot (3 m) ceilings you will want a baseboard that is about 6 inches (15.2 cm) thick. There are a few stock profiles, which vary in height from about 4 inches to 6 inches (10.2 to 15.2 cm), which are available at your local home center. You can also build-up your own baseboard with a base cap or a shoe molding depending on what style you are trying to achieve.

The baseboard typically is a flat board that protects the lower portion of a wall from scuffmarks.

How to Install Baseboard

Tools

- Chalk line
- Pencil
- Baseboard
- Finish nails
- Miter saw (if cutting pieces yourself)

1.
Measure the length of the wall and cut the piece of baseboard trim about ½" (1.3 cm) longer than the wall.

2.
Place small marks on the top of the trim. Cut the board (in this case, the board needs a miter cut for the outside corner exactly where the board extends beyond the corner.)

3.
Once cut, reposition the trim and nail to the wall using finish nails.

4.
Cut another piece of trim with a mitered corner and fit the mitered corners together.

5.
Tack the base shoe molding in place with a finish nail. Continue placing and fastening around the room.

6.
Fill the nail holes with wood putty and sand until smooth.

PART TWO
Design Details

Now that we've discovered the basics of trim, we'll look at how designers and architects create beauty, warmth, and detail through incorporating trim into the living spaces. Rooms without trim are merely unappealing two-dimensional spaces. Trim enlivens spaces and fills them with character, depth, and dimension. Along with these design aspects, trim also defines our living spaces. For example, a column screen can separate a hall from a living room while a built-in colonnade and bookshelf can act as a "wall" to a cozy inglenook. Although trim can help divide spaces, it can also create cohesiveness within the design of a home such as incorporating the same trim profiles in adjacent rooms—hallways, dining rooms, and living rooms. Trim creates balance and rhythm in our homes, making our dwellings harmonious and welcoming.

Trim can transform a room from a mere box to a richly detailed space, creating shadow and depth in our living spaces.

CHAPTER Five

Kitchen and Pantry Trim

Good kitchen design enhances not only the room's utility but also the way it looks and feels. Trim, casings, and built-ins play a large role in that overall aesthetic. When including trim detailing in the kitchen, you might want to consider some material factors. For instance, kitchens are notoriously high-heat and high-humidity areas in the house and when it comes to selecting finishes or materials, choose coatings that clean up easily and can take the high moisture and heat. Consider what style you want your kitchen to be and choose your built-in cabinetry to match this style. Today, kitchen design is moving toward either ultramodern or historical styles such as the scullery. In the latter, designers are using built-in cabinetry that resembles pieces of freestanding furniture. Another popular feature for built-in cabinets is to use only undercounter cabinets, omitting above-counter cabinets altogether. In this chapter, we'll explore the trim design details that make our kitchens more beautiful places to cook, eat, and entertain.

Kitchens have become the center of family life and entertaining, and designers are treating them with high-style design. Shown here is formal built-in cabinetry with box moldings juxtaposed with hand-hewn ceiling beams.

Sunny Disposition

A master of architectural detailing, Benjamin Nutter designed this inviting open kitchen in a Georgian Colonial farmhouse. Located in a northern climate, the house takes advantage of the southern exposure through a wall of five sash windows topped with five clerestories. These windows are trimmed in a simple bed mold casing with a back band, which creates added thickness and weight to the window. Resting on the casing is a bracket, which holds a 14' (4.3 m) -long shelf. Acting as a plate rack, the shelf holds the homeowners' extensive collection of country ceramics. "The bracket detailing is a design element that I picked up from the home's exterior details," says Nutter. To break up the expanse of wall between the two sets of windows, Nutter incorporated a standard stock crown molding. The band of clerestory windows, which Nutter added to allow additional light in the story-and-a-half height space, receive the same casing profile as the windows below. To divide the kitchen and dining area into different spaces, Nutter employed a large rustic beam along the ceiling (not shown). The beam breaks up the spaces visually without the need of adding a wall.

The built-in cabinetry designed by Crown Point Cabinetry has flat-panel pine doors painted in a soft linen color. Raised up on legs, these cabinets are reminiscent of cabinetry from the early 1900s. All the storage is below the counter, a popular way to design kitchens today. A center island with simple trim detailing completes the kitchen's workstation. Although the trim detailing is seemingly simple, it adds just enough texture and shadow to the space to break up an otherwise sparse space.

Design Details

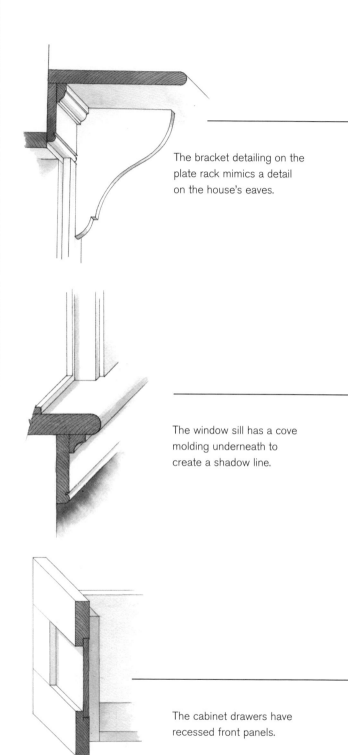

The bracket detailing on the plate rack mimics a detail on the house's eaves.

The window sill has a cove molding underneath to create a shadow line.

The cabinet drawers have recessed front panels.

Under counter built-in cabinets with flat-panel doors and drawers are set up on legs, making the cabinets look like furniture. A plate rail tops the five-bay window casings. The five clerestories have the same apron profile as the lower windows.

Pretty Pantry

Although a service area within the home, this pantry designed by Christine G. H. Franck needed to have a high style—which included creative trim details—to correlate with the rest of the classical house. The ceilings in the room are 10 feet (3 m) high, so Franck incorporated moldings that were heavy enough for the height of the room. Franck advises keeping the trim detailing consistent throughout rooms that open onto one another such as this pantry, which opens onto the dining room. She designed a crown molding from three pieces of trim, and this is the same crown molding found throughout the rest of the first floor's more formal areas—the living room, the hall, and the dining room.

Franck added a beaded casing with a back band and plinth block to the window. Underneath the window, Franck added a box mold. Plinth blocks are great traditional elements, she says. The door casing matches the window casing, offering depth and substance to the space. The casings on the first-floor windows were brought down to the floor rather than simply resting on the sill. This increases the verticality of the proportions of the windows and relates it to the door openings. The head height of the doors and windows is the same—8 feet (2.4 m). The wall between the upper and lower cabinets was finished with bead board, a traditional wall finish for pantries and kitchens. The bead board creates a more informal space amid the high-style classical detailing of the window and door casings. "You do not want just drywall at that surface above the countertop," says Franck. "We used the Georgia Pacific Ply-Bead wood panels. They look like individually set boards without the contractor spending the time setting each board." For added storage, Franck ran the upper cabinets to the crown molding. The bottom of the lower cabinetry is finished in a black paint. "This was a historical way to finish baseboards," says Franck. Black was less likely to show any scuffmarks.

Design Details

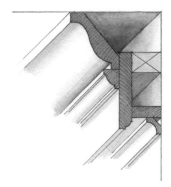

The molding at the ceiling is made up of several pieces, including an ogee, an astragal, and cove profiles.

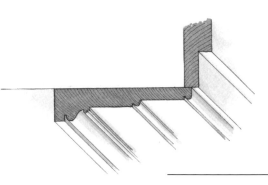

The door casing is stepped with a back band surround.

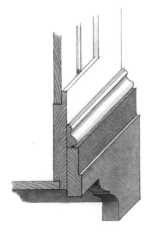

The built-in cabinetry is up on legs to offer the look of a piece of furniture.

Designer Christine G. H. Franck designed this butler's pantry in a coastal cottage. The trimwork replicates the other trim found throughout the first floor.

Classic Cabinetry

With only 7' x 6' (2.1 x 1.8 m) to work within, architects and couple Anne Fairfax and Richard Sammons create a wonderful small kitchen in their city apartment. With everything in reach, Sammons says it is the perfect kitchen. The couple moved the kitchen, which was originally in the front corner of the building, to the back of the house. An eating area complete with a small fireplace sits on one wall and on the opposite wall, a row of custom-designed cabinets. The cabinet detailing reflects the building's Colonial Revival and nautical origins, says Sammons. But most of these details were lost through decades of neglect. The cabinetry is made of poplar with latticed glass on the upper cabinets, inset panels on the lower. The upper cabinets are very decorative. "Most people don't want upper cabinets anymore," says Sammons, "but we chose to incorporate them because we needed the storage space." The back wall is shared with the stairway, leaving a mere 6' (1.8 m) for the cabinets and range and another 6' (1.8 m) for the lower cabinets. The cabinet above the stove looks like a ship's porthole with a keystone, which further extends the home's nautical theme. The dark Honduras mahogany countertops contrast with the creamy white cabinetry. "We always design our cabinets to contrast with counters," says Sammons. They used drawers instead of shelves wherever possible, which is another way to maximize on such limited space. A 5.5" (14 cm) bead covers the entire room. "Smaller bead board from the lumberyard would have looked too busy," says Sammons. The result is a dramatic kitchen that packs a lot of design into a tiny space—just perfect for this active architectural duo.

Design Details

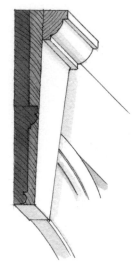

The keystone in the porthole window is detailed with a flat astragal.

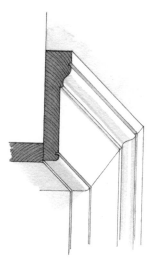

A cove and astragal profile make up the crown.

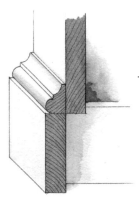

The baseboard is made up of two pieces: a 2' x 4' (0.6 x 1.2 m) board and a stock base cap.

For this city apartment, architectural firm Fairfax and Sammons designed a small kitchen based on Classical Revival forms.

Rustic Kitchen

For this rustic country kitchen, the architect chose to incorporate exposed salvaged beams into the design. Worn by weather and time, the massive beams were rescued from an old barn that was being torn down. The exposed timber shows it had been hand hewn generations ago. Atmosphere and time have also created a mellow patina to the wood. The designer chose to frame the cooking station with support beams and a cross beam. The cathedral ceiling's roof rafters are exposed, offering a further rustic feel to the design. Although the supports are functioning, the rafters are merely decorative. Salvaged wood is one of the best ways to get an old-house feel in new construction. Salvaged lumber is typically first- or second-growth wood, which is a denser, stronger wood. When it comes to salvage materials as well as sources, you have several options from which to choose. Many companies dismantle old barns and factories and sell the reclaimed lumber. Today, distributors are even going overseas to mainland China to seek out sources. Although there are some shipping fees that a salvage dealer might incur, the price points are relatively inexpensive. One word of advice when choosing salvaged wood for trim is to make sure your contractor or salvage supplier smoothes splinters and removes any old nails (typically done by using a metal detector). When using this type of old-growth wood in new construction, also look out for any signs of rot. Adding a bit of the past to the space will be solid, warm, and welcoming.

Design Details

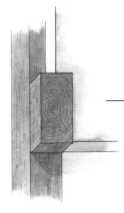

The post-and-beam construction frames the cooking area.

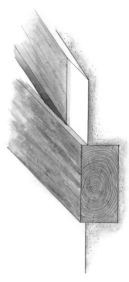

Exposed roof joints rest on an exterior wall plate.

The legs of the center island are tapered.

Rustic hand-hewn beams were salvaged from an old barn to create a new, old-house kitchen in this country farmhouse.

Whimsical Trim

For this contemporary kitchen and pantry area for a young family, the architect wanted to play with the trim elements to create a fanciful and fun design for his clients. Two-tone custom trim work is stained cherry. Whimsy was the word when it came to custom cutting the pieces for the spaces. The door casing is a decorative semicircle with a spiral motif and the reverse of this design is picked up in square columns flanking the islands. Although these scrolls are very contemporary in flavor, the form is reminiscent of an ancient Greek Ionic column capital. The wall treatment is two-toned horizontal 1" (2.5 cm) board—a play on the traditional vertical bead board one would typically find in a kitchen and pantry. In the pantry, the architect chose to incorporate wainscoting reminiscent of a picket fence. This playful treatment of traditional trim elements creates a sunny kitchen space.

Design Details

The decorative scroll work on the column was designed with a custom cut knife.

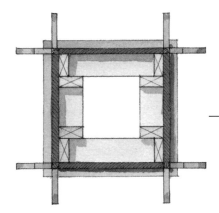

The column in the kitchen is a simple construction of 2' x 4' (0.6 x 1.2 m) boards with a cherry panel. It acts as a space divider between the work area and the pantry.

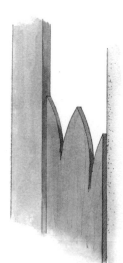

Another whimsical touch is the tulip-shaped wainscoting.

Fanciful trim elements come into play in this kitchen and pantry.

CHAPTER Six

Public Spaces

Areas where trim and moldings can make the most impact in your home are the public spaces, such as the front hall entry, living room, dining room, and family room. These are areas where you'll find the most elaborate trim details in well-designed homes. These elements can help create balance, harmony, and rhythm within the space. Typically, there is a hierarchy of trim throughout the floor plan of a house—with formal, public spaces getting the most attention to trim. Ceilings are adorned with crown moldings, walls are dressed in decorative chair rails, and floors are finished with deep baseboard—while private or service spaces receive simple, unadorned detailing. These public spaces are the places to put your money and effort when it comes to adding trim. Whether your personal taste is ultracontemporary, modern, Arts and Crafts, Neoclassical, Colonial, or Greek Revival, there is a molding profile to match your desired design. In this chapter, we'll look at how architects and designers specify the molding details in public spaces to help transform an empty box into a detailed, inviting space to entertain and live.

A beam ceiling and a simple casing and baseboard make up the elegant trim in this room designed by Krumdieck A + 1.

Farmhouse Classic

An architect by trade, John B. Murray looked for years for the perfect spot to design and build his country retreat house. After searching for just the right piece of land, he came across a dilapidated Greek Revival farmhouse on 100 bucolic acres and he knew it was the place. Although he would not be starting the project from an empty slate, the house needed an overhaul in every way. Murray's challenge was to bring back the integrity of the original detailing while updating the home for his young family. Built in the mid-1800s, the house had very few of its original details left intact.

To restore and enhance the original design, Murray pulled inspiration from the few bits that were remaining—particularly the window casings and paneled bases. The living room needed to accommodate his son Jesse's baby grand piano and be roomy enough for his other two boys to practice the violin and cello, so Murray combined two smaller rooms and incorporated a fireplace as the centerpiece. Playing off the original window base paneling, Murray replicated that trim profile for the fireplace surround and added a simple mantelshelf with Greek Revival accents—an appropriate design for the age of the house. The window casings were embellished with a dog-eared molding—a typical design feature in Greek Revival homes in the region. To keep balance and harmony within the design, Murray minimized the size of the crown molding. The ceiling height—a mere 8' (2.4 m)—would not accommodate a deeper profile. Built-in cabinet doors at the far end of the room convey the same wall-frame molding found under the windows. Above the mantelpiece, Murray incorporated another wall-frame molding, which ties the design together beautifully.

Design Details

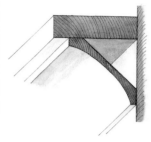

The mantel is made up of a simple shelf and a cove molding.

The crown molding consists of a reverse ogee between two bands.

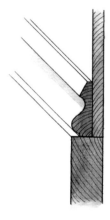

This wall-frame molding has a reverse ogee with a beveled band and is used above and below the mantelshelf as well as under the window.

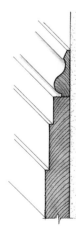

The baseboard is a simple stepped molding, topped with a reverse ogee.

Architect John B. Murray designed Greek Revival-style trim for this farmhouse. Note the dog-eared trim around the window casing.

Greek Revival Living

When the architectural firm Historical Concepts set out to design this grand, "new old house," they looked to the past for inspiration. Historical Concepts chose the design vocabulary of the Greek Revival antebellum home, researching pattern books of the early 1800s to re-create the classical details within the room. Author Russell Versaci explains pattern books of the past were the only learning tools carpenters had at their disposal to build houses. Today, they lend a hand to craftsmen creating past forms. Once fully immersed in the language of the style, Historical Concepts chose to create a high-style interior for this formal living room. Although the open layout is modern—a traditional Greek Revival house would have had a stair hall in the center of the house—the detailing is true to its style. The entryway is classic Greek Revival architecture with its temple front made up of a pilaster jamb, frieze, and cornice, topped by a pediment. The details are scaled to match the large room. The floor-length windows are trimmed in a wide, tapered casing in the Greek Revival style. The interior doors also have the same wide casing. The tops of the doors, along with the window casings, have dog-eared detailing around the transoms. The transoms are a clever way to create airflow within the house. The ornamental crown molding and baseboard follow the elaborate trim scheme in the room. The mantel is also a classical Greek Revival design and plays off the other trim details in the gracious updated antebellum.

Design Details

The flat frieze and cornice are made up of several pieces. Note how the simple cove moldings and planes built up can create a beautiful composition.

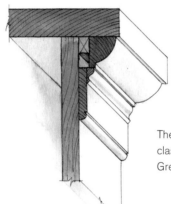

The pilaster door trim is a classic example of the Greek Revival style.

The 6" (15.2 cm) -wide taped casings around the doors are dog eared.

Historical Concepts used the vocabulary of the Greek Revival style to create the elaborate trim for this room.

Classical Style

When Jeffery L. Davis wanted to build a home right on the beach, he decided that a classical villa modeled after ancient Greek architecture would be a fitting design. Davis hired designer Christine G. H. Franck to bring his vision to fruition. Franck is well versed in the classical language of architecture and she designed a temple front exterior with four columns. One goal within the interiors was to use stock trim items where possible. "This was a challenge because today's stock trim is rarely beefy enough for classical profiles," says Franck. Davis wanted traditional formal spaces but also wanted these spaces to be a part of an open-floor plan. The house's interior is not broken into rooms by conventional walls but with column screens instead. The columns are of the Greek Ionic order. The openings are made up of the entablature. Holding up the entablature is the column and next to the column is the ante, which is the terminating feature on the wall. The room's ceilings are 10' (3 m) -high so Franck wanted the molding profiles to accommodate that room height. Everything in the room is generated from the columns and entablature, says Franck, "and the crown molding is created on the principles of the Ionic order." Door and window openings have a wide, beaded casing and back band to give them thickness equal to about ⅙ of the width of the opening. All wood trim is poplar.

Franck chose to paint all the trim in white except for the plinth of the baseboard, which is painted black to give contrast in the room and help hide wear and tear. The living room mantelpiece is based on popular American and English eighteenth-century designs, particularly those of Abraham Swan. The mantel has stock corbels. "And a molding was cut off the top of the corbel so that we could integrate it better into the mantelshelf and the bed mold for the mantel shelf," says Franck. Although Franck had to stay within budget and the goal of using stock material, the room is full of classical elements that work in perfect harmony.

Design Details

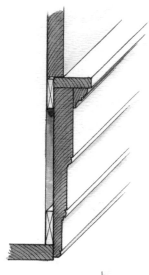

The entablature is made up of stepped trim.

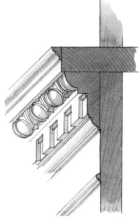

The capital of the ante had egg and dart detailing as well as dentil molding. Note how the piece is built up by using four pieces.

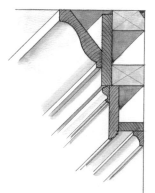

The crown molding within the room is made up of six pieces. Note the cove molding at the top of the piece.

Designer Christine G. H. Franck used a combination of stock materials and custom trim for this classically styled living room in a coastal cottage. Note the fluted column screen, which acts as a room divider.

Arts and Crafts-Inspired Inglenook

Inglenooks and colonnades, or room dividers, were popular design elements in Arts and Crafts houses at the turn of the last century. Economy of space was a key element in small bungalows and Craftsman houses, and the built-in offered space-saving alternatives to heavy freestanding furniture. For the design on this contemporary Arts and Crafts-inspired home, the architect relied on forms established in the American Arts and Crafts movement, which was influenced greatly by Japanese architecture. The focal point of the living room is a massive stone fireplace complete with an inglenook. Colonnades with square, tapered columns act as a room divider and create the cozy spot in front of the fireplace. To break up the massive expanse of wall space in the room, the architect incorporated two-tone horizontal boards in a warm honey separated by a thin band with a deep cherry finish. The tapered column is picked up in the staircase newel post. The architect chose to further break up the space with a series of windows that flood the space with natural light. The window casings are simple flat board casings. Ceiling beams are exposed to offer a lodge feeling in the design.

Design Details

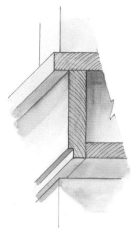

Horizontal and vertical boards create a dramatic effect.

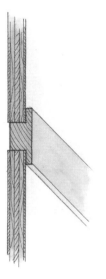

Horizontal fir strips create interest on the wall paneling.

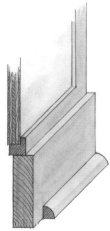

The baseboard has a simple base shoe at the floor.

A massive stone chimney is the focal point of this dramatic Arts and Crafts room. A colonnade with tapered columns divides the inglenook from the main living space.

Asian Influence

In this ultracontemporary condominium dining room, designer Fanny Haim chose to incorporate an Asian flair into the design. Symmetrical doors reminiscent of traditional Japanese Shoji screens (which are made of translucent rice paper) are a major element in the design of the room. Truly a tranquil setting, the room's color scheme is divided in two. On one side, the walls are cream and have 3' (0.9 m) -wide horizontal bands separated with a thin band. The treatment is slightly reminiscent of ancient Greek rustication or medieval board-and-batten construction and creates a look that is strong and bold. A wall niche is incorporated into the sleek design. On the opposite wall is the same banding but done in a deep cherry wood. The contrasting wall colors add visual interest and sophistication to the space. One wall panel is recessed and is actually a sliding door to a china cupboard. Although the design is informed by ancient architectural elements, the dining room is every bit a modern design and offers a tranquil spot to have a quiet dinner.

Design Details

The walls receive an ogee molding to add depth and character.

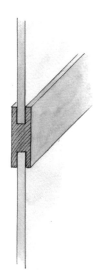

The shoji screens have decorative horizontal banding.

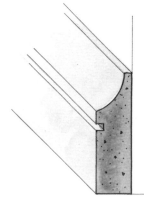

The baseboard is a stock piece without a base shoe.

This room has a contemporary feel and is inspired by Japanese design.

chapter Seven

Private Spaces

Whether your private space is a luxurious, contemporary master bath or a cozy, country guest room, there is trim detailing to match your style. One design element to consider is the hierarchy of trim within the house. Oftentimes, the most elaborate trim is reserved for the formal public spaces of the home—the living room, the dining room, and the front hall foyer. Private spaces are typically smaller than more formal rooms so trim-work should be proportioned accordingly. For instance, if you have 10' (3 m) ceilings on the first floor of your home and 8' (2.4 m) ceilings on the second floor, you'll opt to use a chunkier, more decorative molding on the first floor. On the second floor, reserved for private spaces such as bedrooms and baths, you will most often opt for a standard stock molding. Private spaces such as service areas (butler's pantries, kitchens, mudrooms, and laundry rooms) often receive a simple 1' x 4' (0.3 x 1.2 m) paint-grade, flat-board trim casing around doors and windows. Another popular trim detail in these utilitarian spaces is bead board. Bead board is much easier to clean than straight drywall and adds an old-time quality to the space. Bead board comes in a variety of widths to coincide with the proportions of the space. Another factor to consider in private spaces that produce a lot of humidity is the finish you use on your trimwork. High-gloss paint is standard fare in these rooms. Remember that it is in these private spaces where you might be able to save some money on your trim projects. Put your money into the spaces that you will use and enjoy most often, as opposed to small areas that you visit infrequently.

Built-ins are often incorporated into houses as space savers. They may include hidden storage as an added bonus.

Beautiful Bath

When it comes to adding a bathroom to an old house, it is often hard to decide where to fit one in and, oftentimes, you must forfeit an existing room to incorporate a bath. Instead of looking at the negative points of losing space, look at the positive prospects of the ultimate project. When the room has handsome wainscoting installed, you will have created an attractive sanctuary. These home-owners did this in their Colonial Revival home. Old World charm is found in this pretty bathroom complete with a fire-engine red claw-foot tub. The room looks more like a living space than it does a bathroom and the key to this success is the use of wainscoting and wallpaper. To balance the color in the room, the designer chose to incorporate 3' (0.9 m) -high wainscoting with a box panel molding. The paneling is finished in a high-gloss paint that is easy to wipe clean. A 4" (10.2 cm) -high baseboard is finished in the same white high-gloss paint. The designer also incorporated a wall of built-in drawers to store linens and bath towels. The look within the bathroom is clean, and elegant, and works with the rest of the home's spaces.

Design Details

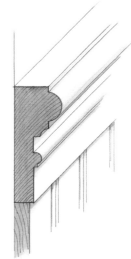

The top of the wainscoting has a shallow, flat-bed molding.

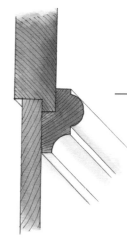

The wall panel is made up of a simple box molding.

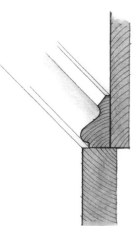

The baseboard is made of an ogee molding on top of a flat board.

The designer of this pretty bath incorporated wainscoting and built-in cabinetry painted a crisp white that contrasts with the red tub.

High-Style Bath

Drama abounds in this master bath design by architectural firm Historical Concepts. Elements of the Greek Revival style fill the space, including the custom trim work. Historical Concepts created a lavish master bath with 13' (4 m) ceilings based on historical profiles of antique Greek Revival trim work found in houses built between 1820 and 1840. A high-style window casing frames the claw-foot tub, making it a highlight within the room. The casing is made of a built-up entablature, pilaster, and box-frame detailing. The same detailing is picked up in the door surround. The mirrors are framed in moldings that replicate a traditional dog-eared detail prevalent throughout antique high-style Greek Revival houses; this molding detail is also found downstairs in the more formal areas of the home. The simple flat-board baseboard is 7" (17.8 cm) thick. Because the space is private, the detailing is not as pronounced in this area as it is in the more formal rooms found on the first floor, although this space is much more elaborate than many bathrooms and creates a serene atmosphere.

Design Details

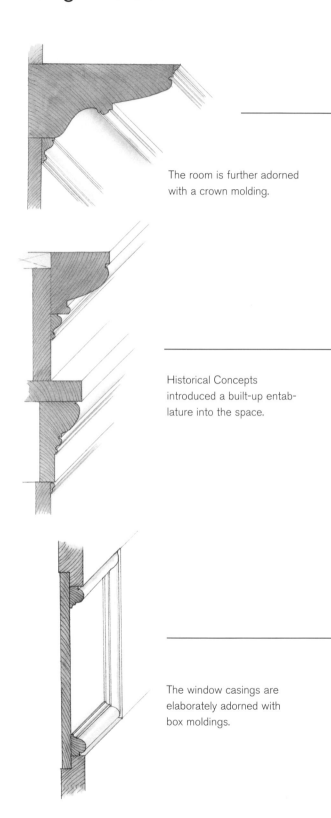

The room is further adorned with a crown molding.

Historical Concepts introduced a built-up entablature into the space.

The window casings are elaborately adorned with box moldings.

This bath's trim elements are based on Greek Revival forms to create a dramatic space to soak in a hot tub.

Classic Vanity

The master bath in this coastal cottage designed by Christine G. H. Franck is compartmentalized into three chambers: the vanity, the water closet, and the soaking tub. The separate rooms are painted a crisp white, and the same molding details are used throughout to create continuity within the space. The vanity top is a cool Carrara marble. The architrave on the master bedroom side of the entry door is 5" (12.7 cm) tall, including its stock back band and a custom-milled casing. On the inside of the master bathroom door, the molding is 4⅜" (11.2 cm). "Both the casing and back band are custom milled," says Franck, "though both are based on stock profiles."

Hierarchy is important when it comes to outfitting spaces with trimwork. "On the second floor of this house, the cornice height will differ depending on where it is used," says Franck. "For instance, the cornice in the master bedroom is 5⅝" (14.3 cm) high—plus a bit for the bed mold. It is custom and inspired by the Tower of the Winds (an ancient tower in Athens). In the master bathroom, the cornice is 3¹³⁄₁₆" (8.4 cm) high and is a stock crown molding. It is one piece consisting of a cyma recta, fillet, and cove molding.

The wainscot at the vanity are panels installed directly on top of the water-resistant wallboard surface. The wainscot's wood stiles and rails are ¼" (6 mm) thick with a cap molding at the top of the wainscot. "This is not as fine a detail as, for example, running the drywall only to the top of the wainscot and then building true panels of stiles and rails and bead board in the field, but it is an easier, faster, and less expensive alternative, giving a very similar look to a true panel once it is painted," says Franck.

Design Details

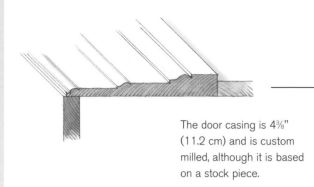

The door casing is 4⅜" (11.2 cm) and is custom milled, although it is based on a stock piece.

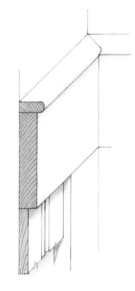

The bead board wainscoting is topped with a shallow lip or cap molding.

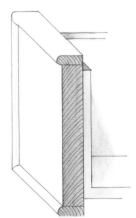

The built-in cabinetry has a box molding motif.

Designer Christine G. H. Franck designed a bathroom with a simple trim details.

Barn Bedroom

This pretty bedroom is an addition to an old country farmhouse. Architect Bruce Norelius of Elliott Elliott Norelius Architecture says about the project: "There is something about the way the house sits on the land, the simplicity of it. You recognize it as a special house." This includes not only the handsome, spacious addition to the original 1830 structure but also the design details chosen during the design process. The master bedroom was incorporated into the new barn addition on the back of the house. The architects wanted the structure to appear as if it had grown organically over time and a barn-like structure would achieve this goal. Not only does the exterior of the structure have elements of a barn, but the interiors do also. The structure's king-post truss (made of a series of members connected in a web or triangle to create a rigid structure) was left exposed as in an old barn. The pine trusses function to support the roof as well as add a design element to the space. Horizontal flush boards cover the walls and are painted white, which contrasts with the honey-colored pine ceiling and truss. The windows are merely cut into the boards, furthering the rustic, sparse barn feel to the room. A 7' (2.1 m) free-standing wall of bead board separates the master bedroom from the master bath. The lower beams or panels of the truss rest on this wall. "I think the barn addition has two readings," says Norelius. "When you walk in, it feels natural and comfortable. It's also a little stripped down beyond what would feel typical and feels a little more architectural." The room is not just a solution to a problem; it is real architecture. The "bedroom in a barn" is compatible with the rest of the historical structure.

Design Details

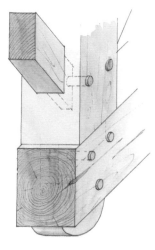

The roof's truss system is exposed. Note the peg joinery.

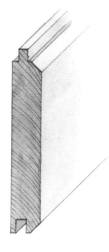

The walls are made up of a series of 5" (12.7 cm) horizontal boards.

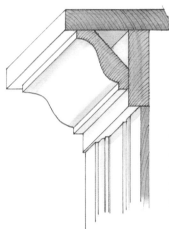

Behind the bed is a wall to divide the master bathroom. The wall is dressed in bead board and has a stock crown molding.

The exposed roof truss in this master bedroom addition is reminiscent of barn structures in this rural area.

Wonderful Windows

For this Greek Revival house, historically appropriate moldings were a key design element. Architectural firm Historical Concepts relied on old building pattern books from the 1900s. "What you notice first is the light within the room," says principal Jim Strickland. A wall of windows overlooks a river, three more overlook the porch, and two on either side of the bed create a room that is seemingly all windows. Historical Concepts chose to frame the windows with pilasters and follow the Doric order of classical architecture. Although the master bedroom is just off the master bath, the bedroom's ceilings are a full 4' (1.2 m) shorter than the bathroom ceiling. There is continuity between the two rooms' molding treatments but because of the differing ceiling height, the bedroom's moldings are a bit more pared down. The built-up custom molding above the windows to the ceiling consists of an architrave and a crown. These are correctly proportioned for the space. The use of the Greek pilasters as frames makes a dramatic effect within the space. The moldings are painted a cream color and contrast against the soft green walls. The room with views from three sides not only is sunny and bright but also offers a wonderful frame of high-style design within the space.

Design Details

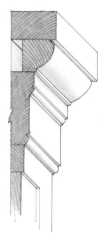

The windows are framed with pilasters.

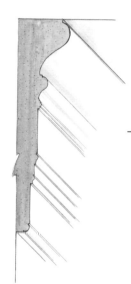

The elaborate crown molding tops the frame of the windows.

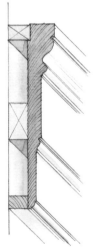

The window trim is carried between the windows and is a custom-stepped design.

Historical Concepts framed this bedroom's windows with pilasters for a high-style sleeping chamber.

CHAPTER Eight

Built-Ins

Built-ins have been popular since the turn of the century; inglenooks, window benches, colonnades, cabinetry, open shelves, and breakfast nooks even Murphy beds have been working into the design of houses seamlessly. First and foremost, built-ins offer economy of space within the home. In the compact floor plans of many kit houses, bungalows, and Craftsman houses in the early twentieth century, built-ins created valuable floor space as well as valuable storage space. Built-ins often incorporate many of the trim elements found within the home. Colonnades offer double duty not only as bookshelves but also as room dividers. The colonnades' columns should always reflect the home style. For instance, if it is an Arts and Crafts house, the columns will be square and tapered, while a Colonial Revival house might have a colonnade with a squat Doric column. Window seats offer extra storage and seating, and the bench often will include trim features found in the wainscoting. Cabinetry also picks up the same trim design in a room—be it a crown molding profile or a baseboard width. Remember that when adding built-ins, continuity is the key element in the design. If you have several built-ins within one space, make sure they get the same design elements in the way of moldings and trim, and your rooms will add not only wonderful permanent furnishings but also style and grace to the design.

Built-ins not only add a wonderful design element to a space but also are great space savers in small houses.

Breakfast Nook

For a major renovation to an old house—that was literally ready for the wrecking ball—architectural firm Rill & Decker came up with creative elements for redesigning the space for a growing family. The house was an 1890 Victorian that had sat vacant for decades. Windows were boarded up and there were gaping holes in the ceiling. "We needed to reconfigure all the interior spaces that no longer functioned for a modern family," says principal architect Anne Decker. These spaces included a new kitchen. "Economy was a factor when designing these rooms," says Decker. "We went with a clean, relaxed farmhouse feel within the space." The team decided to include a breakfast nook for its warm cozy appeal. "It was our recommendation to the homeowner, based on economy of space and wanting to add the warmth of sitting in a cozy booth," says Decker.

The benches are made of paint-grade birch veneer plywood with decorative nosing, and bead board is set within a 3" (7.6 cm) frame. For continuity within the design of the room, Decker chose feet on the benches that are similar to the ones on the cabinets. The benches are finished in a semigloss paint. Decker recommends a catalytic paint finish for its durability in kitchens and baths. The built-in cabinets are made with flat-panel inset doors with an oval surround. The door stiles and rails are 2¼" (5.7 cm) with concealed hinges to create a flush exterior—again a design choice made with economy in mind. Bead board is used in the hutch-like piece to tie back into the nook details, says Decker. "Bead board is easy to clean and is durable. Drywall is much more unforgiving—especially when small children are around."

Trimwork in kitchen areas is best kept simple for cleaning purposes—at least near the sink and cooktop. Decker also advises foregoing a wooden apron under a window placed above a sink. "You should opt for either tile, granite, or a marble backsplash, again for maintenance purposes."

Design Details

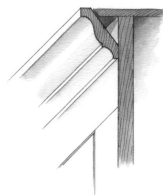

The top of the cabinetry has a stock crown molding piece for a finished look. Note that the ceiling has no molding.

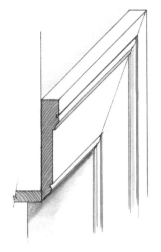

The window casing is fitted with a mitered joint.

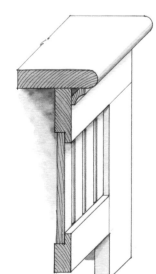

The built-in bench picks up the bead board detail that is found on the wall above the counter.

Architectural firm Rill & Decker added a breakfast nook to this country-style kitchen.

Reading Niche

Architect John Cole was inspired by Arts and Crafts design for a custom-built coastal camp cottage. The homeowners wanted a house that looked as if it had always been there, so he built a dwelling of rough-sawed cedar—befitting the woodland surroundings. "We wanted the exterior character to come through in the interiors so we incorporated reclaimed Douglas fir lumber from an old Navy shipyard into the interiors," says Cole. "The homeowners also requested that we eliminate the need for lots of furniture so we incorporated built-ins into the house—benches, dressers, drawers, and shelving." Cole found creative ways to work in built-ins wherever he could. In the guest bedroom, he added a niche bookshelf. "The angle of the bottom of the shelf follows the slope of the eave," says Cole. The back of the bookshelf is finished in the same bead board as the wainscoting, which is 3' 6" (1.1 m) high—the boards are no more than 1½" (0.5 m) thick. The top rail of the wainscoting is 5½" x 4" (14 x 10.2 cm). The window and door casings are simple flat board casings also in Douglas fir and are also 5½" x 4" (14 x 10.2 cm). This casts a nice shadow, says Cole. The ceiling is ½" (1.3 cm) pine tongue-and-groove board, a typical treatment in coastal cottages. The trim and the built-in niche in the room create a warm and cozy atmosphere in this camp cnvironment.

Design Details

Bead board is not just for walls but is a popular finish for rustic ceilings as well.

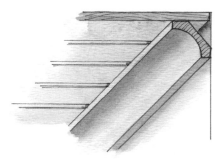

A stock cove molding at the ceiling offers a finished look.

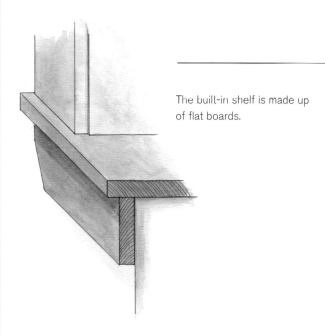

The built-in shelf is made up of flat boards.

This simple bookshelf designed by architect John Cole is cut at an angle at the bottom to follow the house's roofline.

Shaker Simplicity

For a custom Georgian Colonial farmhouse, architect Benjamin Nutter informed his design with the Shaker style. He created rooms that are sparse and simplistic. "Economy of space is a key feature in Shaker design," says Nutter, so he incorporated the furniture right into the design. Two built-in wardrobes flank a three-bay window and window seat. The owners like their home to be organized and they wanted loads of built-in storage to keep them that way. "The belt crown molding, which is set at the top of the windows, is built into the top of the wardrobe," explains Nutter. This offers continuity in the design. The 6" (15.2 cm) baseboard with a simple bed mold also plays into the bottom of the wardrobe and window seat. The window seat itself offers ample storage for bedding—comforters and extra blankets on cold northern nights.

Design Details

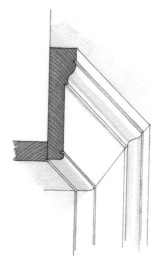

The window casing has standard miter joints.

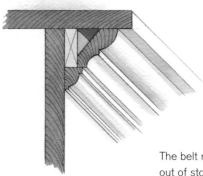

The belt molding is made out of stock crown and flat astragal profile.

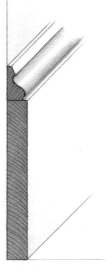

The baseboard is 6" (15.2 cm) thick to accommodate the ceiling height. The board is topped with a base cap.

Architect Ben Nutter designed a wonderful, bright bedroom with a built-in wardrobe and window seat. The design is tied together through the use of a belt crown molding.

Built-In Hoosier

One of the hottest trends in kitchen and pantry design today is to create built-in cabinetry that looks like free-standing furniture. In this contemporary country farmhouse, Crown Point Cabinetry did just that. This built-in cabinet is made to look like an antique hutch or Hoosier cabinet of the early 1900s. Upper cabinetry has glass-front doors with three shelves for dishes. The back of the counter is finished in bead board to give it a country feel. The piece is set up on legs, again another popular feature in antique cabinetry. The doors are flat paneled with a square inset and round wooden knobs. The top of the hutch is finished with a decorative crown molding; the same crown molding profile is used on the other built-in cabinetry within the room.

Design Details

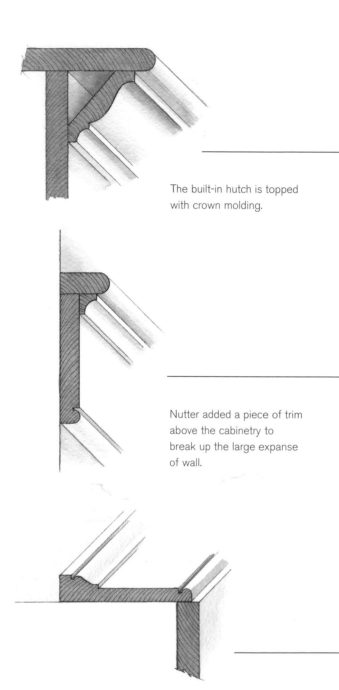

The built-in hutch is topped with crown molding.

Nutter added a piece of trim above the cabinetry to break up the large expanse of wall.

The door casing is a simple stock profile.

Hoosier cabinets were popular at the turn of the last century; today, custom cabinet makers are bringing back the look into the kitchen and pantry.

Camp Cottage

Architectural firm Bernhard & Priestley strive to create houses that fit into their landscape and social context. Architectural history is a big part of that design philosophy. In this renovation of a shingle-style home in a woodland coastal community, the architects wanted to create a rustic environment in the interior as was on the exterior. "Natural materials are crucial to the design like this," says Richard Priestley. The architects chose pine interior trimwork in the way of bead board, wide casings, and a built-in bookshelf for a hallway. The pine is finished in a clear coat of varnish. The narrow, short bead boards are placed between the window casings. The dimensions of the window casing match the dimensions of the center window in the bay. The U-shaped window seat offers a cozy spot to read as well as storage under the seat. The rustic trim is carried through the rest of the house, offering a warm, welcoming vacation home for the owners to retreat to every summer.

Design Details

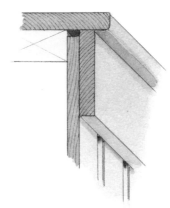

Trim is left simple—merely flat board casings.

The walls in the hall are pine bead board.

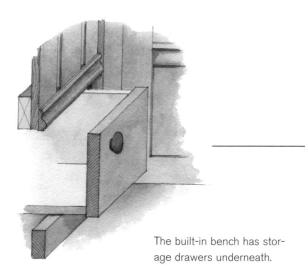

The built-in bench has storage drawers underneath.

Bernhard & Priestley Architecture incorporated wonderful interior elements into the renovation of an old 1960s ranch house, including a U-shaped window bench and a bookshelf.

Staircases

Staircases not only offer a practical way to get between your home's different floors but also are an integral design element within the home—truly a focal point for the home. Whether a grand front staircase or a back hall stair, the design should always work into the architectural theme of the home.

There are as many stair styles as there are home styles, so your home design will determine your stair design. Today, many homes draw on the tradition of classical architecture to create neoclassical styles. The neoclassical incorporates balance and harmony of design, rather than strictly copying a historical style. There are also ultramodern staircases that offer simple treads and risers without tradition handrail balusters and newel posts.

If you are working with a clean slate (new construction), placing the staircase in your home will be an easy task; if you are renovating your home, moving a staircase might be an expensive challenge. Stairs also take up space within the home and are a great impact on the overall floor plan—probably the greatest. If you are planning to replace a staircase, work with an architect before ripping out the old and incorporating the new. Also check your local building codes to determine what the stair regulations are in your state and country.

Staircases come in all shapes, styles, and sizes. If you are planning to incorporate a staircase into your floor plan, make sure that you have the room to do so. Staircases are quite complex and best left to a professional to design and install.

Historical Replication

For this beach cottage staircase, designer Christine G. H. Franck drew inspiration from a historical home in the area. "The stair is meant to be simple, graceful, and elegant—a contrast to the heaviness of the rusticated base that begins outside the building and extends into the ground level of the entry hall," says Franck. "To achieve the delicate lightness of the stair and the simplicity of the design, thin rectangular pickets are used in lieu of turned balusters—the newel posts are also rectangular," says Franck. "The railing is a simple, round profile meant to fit perfectly into the hand." Franck chose to use stock parts for the construction of the stairs. The only compromise in the design of the stairs was the goosenecks and newel caps, which were not as complex or as subtle as historical or custom examples. "The stair also has brackets at each riser, which visibly support the tread," says Franck. These brackets were custom cut on site.

The stair hall is designed to move gracefully from the ground to the main floor and above. The stair hall windows and their tight integration with the landings make the stair seem to float in the hall, which increases the sense of lightness. The walls are rusticated, a construction element more related to the ancient Roman building traditions of Andrea Palladio and the broad classical tradition. "This treatment is found in many Georgian and Federal period houses," says Franck. "The builders wanted the look of a stone rusticated base, or wall, or quoins, or voussoirs, but stone was either not available, or too expensive, or stone craftsmen just weren't here." Rustication seems to be mostly a Renaissance development that gives the base of a building a look of (if not really) solidity and defense. "It made sense to me, knowing I wanted to treat the base with only skirting boards, to use those and transform them into the idea of a protected, solid base, something desirable in such a vulnerable location," says Franck. It also made sense to connect to the American tradition of attempting to render in wood what once had been in stone. She too believes it is simply a good way to articulate an otherwise plain wall surface from ground to first floor.

Design Details

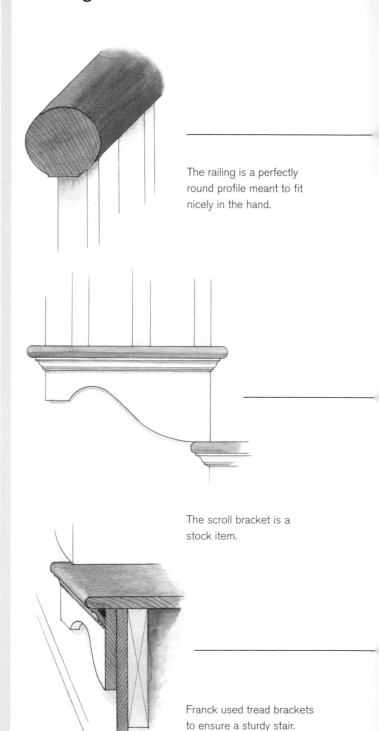

The railing is a perfectly round profile meant to fit nicely in the hand.

The scroll bracket is a stock item.

Franck used tread brackets to ensure a sturdy stair.

Christine G. H. Franck was inspired by a historical design for the staircase in this new old house. Note the rusticated wall treatment, another reference to the past within this space.

Contemporary Railing

Designed by architect Richard Burt, this contemporary vacation house is made of wood, stone, and sandblasted concrete and is designed as a permanent version of a tent on the shore for a small family. "In this regard, the design includes high ceilings of cedar, floors of slate, many wood windows, all combined through an attention to detail and craftsmanship similar to what is found in quality wooden boat construction," says Burt. The sandblasted concrete walls are one of the major design features in the living room. These walls divide both functional areas and floor levels of the house. These concrete walls form a fireplace, a corner table, and even couch supports.

The stair is an important component of this experience because it bisects the main concrete wall and extends the main walkway from the upper levels of the house to the living area. The cherry complements treads and risers made of cherry wood. The unique shape of the handgrip feels comfortable to hold. The vertical supports are detailed in a manner that complements the detailing found elsewhere. "In short," the architect explains, "the railing is an extension of the design concept and the overall feeling of the house."

Design Details

The handrail is a custom design made of cherry wood and is meant to feel comfortable in the hand.

The construction of the baluster, although simple, is sturdy.

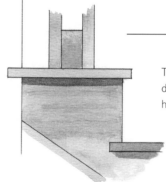

The balusters follow other design elements in the house.

This railing designed by Richard Burt Architects is made of cherry wood and is in keeping with the contemporary design of the house.

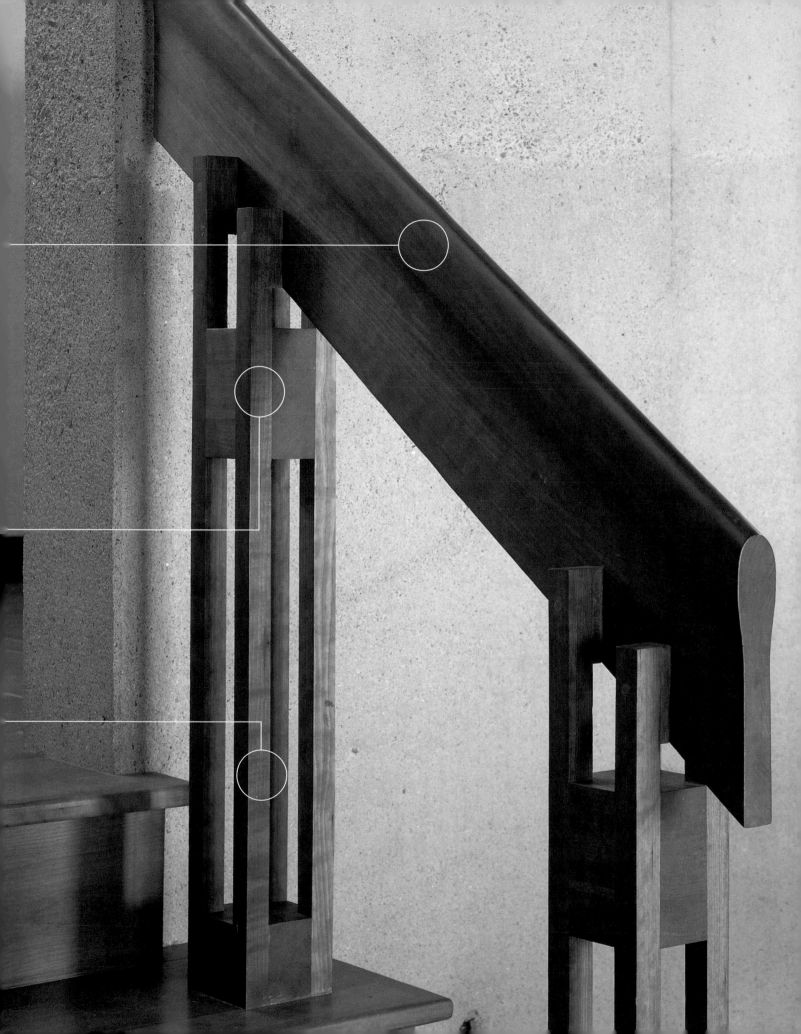

Beach House Hall

The Taylor guesthouse by Polhemus Savery DaSilva Architects and Builders is a wonderful casual spot for visitors to come and stay—away from the hustle and bustle of the main house. "We love looking at historical architecture of all types and we work in an area where nautical influence is prevalent," says John DaSilva. This stair design combines those two things. "Twisted and spiral forms occur in Baroque and Mannerist architecture, such as the twisted columns of Bernini's Baldacchino at St. Peter's, and in the decorations of the Manueline period in Portugal, when Portuguese explorers were roaming the world in ships," says DaSilva. "Many architects used nautical motifs, including rope to celebrate the oceangoing adventures of their time in the built form."

Newel posts are 3½" (8.9 cm) -square posts with shallow pyramidal tops. The balustrade is 3½" x ¾" (8.9 x 1.9 cm) flat boards cut into a flattened "rope" shape. The railing has a 2" (5.1 cm) diameter cylindrical shape with a squared-off flattened sub-rail.

The balustrades and newel posts are poplar and the handrail is oak while the balustrades and newels are painted, there is varnish on the handrail. "We reference historical forms and are inventive within the tradition," says DaSilva. "'Evolutionary rather than revolutionary,' in Robert Venturi's words—this makes for work that can have serious intellectual intent but also be playful, friendly, and evocative."

Design Details

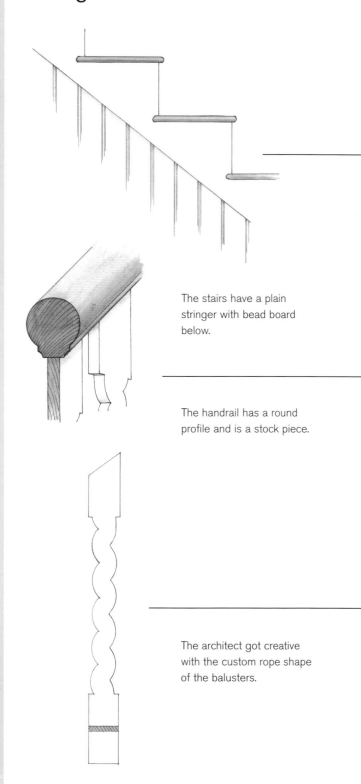

The stairs have a plain stringer with bead board below.

The handrail has a round profile and is a stock piece.

The architect got creative with the custom rope shape of the balusters.

Architect John DaSilva wanted this beach house to carry the nautical theme wherever possible so the balusters on this stair rail are reminiscent of nautical ropes.

Cape Cottage

For this classic cottage's stair hall on a coastal island, architect Mark Hutker wanted to achieve an open, inviting ascent to the second floor. Windows are located at the upper level to allow light to stream onto the stairway all day. Hutker created a straightforward design that enhances the daily event of using the staircase. "Our firm used honest and pure detailing in a logical fashion as if a carpenter had simply used what he had at hand in the age when the Cape cottage was born," says Hutker. Joints are thought out to respect each element of the stairway on its own merit. "Discerning what each element adds to the composition is easy, like the ethics of a good wooden boat. There is nothing used that is not necessary to the overall function of the stair," he says.

Hutker chose to emphasize the strongest elements of the stair. On a practical point, the natural wood is used where handprints would likely show on a painted surface. The newels and handrails are pine with painted pine balusters. The treads are oak with painted pine risers and there are painted pine stringers. Hutker used a primer and three finish coats of oil-based paint applied with a brush, as opposed to a roller, for a wonderful, old house feeling.

The tongue-and-groove wallboard is a 1" x 6" (2.5 x 15.2 cm) V-groove in a clear pine. The boards are vertically oriented to enhance the vertical feeling of the stair space. Simple flat board casings are used for windows and doors. "We felt we had just enough different wood surfaces and finishes that added detail but would not be lost or complicate the simplicity of the design," says Hutker. "We wanted to do what a carpenter might have done at the turn of the last century. The design is simple, plain, and honest."

Design Details

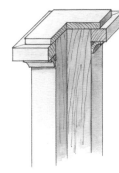

The square newel post picks up the detailing of the balusters.

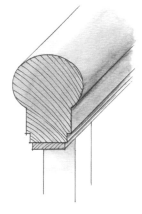

The oval handrail is a stock piece.

The treads are left in a natural wood tone while the risers are painted a crisp white.

This sweet stair hall in a seacoast cottage has a simple, straightforward staircase. "We wanted the space to look like it has always been there," says Hutker.

Eclectic Colonial

After designing 25 houses in a neotraditional beach community, architect Eric Watson decided it was time to design his own home there. His inspiration came from Dutch Colonial, French Colonial, and Mission styles. Intermingling these design elements was quite a challenge, but Watson pulled it off effortlessly. The stair hall is one aspect of this combination of designs. The entry foyer is nestled between two guest rooms and is "orchestrated to heighten the experience of arrival in the second-floor living room culminating in the park vista," says Watson. A Creole-influenced tapered newel post and Shaker-style tapered spindles made a simple yet elegant stair rail. The treads and risers are stained a dark oak to match the floors on the second floor while contrasting with the stone tile floor in the foyer. The newels and spindles or balusters are painted black. The walls have horizontal 1" x 6" (2.5 x 15.2 cm) boards with a simple bead. The staircase is a switchback configuration, which means a second flight turns 180 degrees from the first staircase, typically turning at the landing. Watson creates harmony between historical styles in this entry hall.

Design Details

The handrail has a low, squat profile.

The spindles are tapered in the Shaker style.

The newel post is painted a rich ebony.

Architect Eric Watson was inspired by the Mission style for this staircase in his Colonial-style house. Note the tapered newel post.

Glossary of Trim Terms

A

Abacus: The flat stone slab underneath the entablature that forms the top of the capital of a classical column supporting a beam.

Acanthus: A stylized leaf motif; one of the primary decorative elements of classical architecture. With its origins in Greece, the acanthus was adopted by Romans and transmitted into the general classical tradition.

Aedicule: A term now applied to the frames surrounding a classical doorway or window flanked by a pair of columns and topped by a pediment but which has its origins in the architectural treatment of the shrines of the classical period.

Align: The faces of objects that are in line with each other, or when their center lines lie on the same axis.

Ante: The terminating feature—typically a square or rounded column—on a wall, which supports the entablature.

Apron: The horizontal member directly beneath the stool or inside sill of a window.

Arcade: A series of arches supported by piers or columns.

Arch: A curved structure used as a support over an open space, as in a doorway. A semicircular opening in a wall, or a freestanding structure dependent for its structural stability on the horizontal load threatening to push it apart. Usually made from cut stone blocks forming interlocking wedges.

Architrave: Originally a simple, flat, structural lintel spanning an opening in a wall, it is the lowest part of the classical entablature. Subsequently a term used to describe any molded door or window frame.

Archivolt: One of a series of concentric moldings on a Romanesque or a Gothic arch.

Areaway: The excavated area between the area wall and the basement window.

Arts and Crafts: Galvanized by William Morris, a group of architects and designers attempted to revive the traditions of simple handicraft techniques in nineteenth-century England. In architecture, they looked at the unself-conscious vernacular tradition of barns, mills, and cottages as an inspiration and at the aesthetics of the medieval period. Known as the Arts and Crafts movement, this design tendency spread across much of Europe to America and Australia.

Awning Window: A window hinged along the top edge.

B

Baluster: Any of the small posts that make up a railing, as in a staircase; may be plain, turned, or pierced.

Balustrade: The combination of a railing held up by balusters.

Base Cap: Used in conjunction with the base mold; sits on top of the base mold and runs horizontally with it.

Baseboard: Finish trim where the floor and walls meet.

Base Molding: The decorative wooden strip along the top edge of the baseboard.

Bas-relief: A low-relief sculpture that projects only slightly from its two-dimensional background.

Base Shoe: The wooden strip (usually quarter round) along the bottom face of the baseboard at the floor level.

Batten Board: A small strip of wood used, for example, to cover the joints between vertical siding.

Bay Window: A set of two or more windows that protrude from the wall. The window is moved away from the wall to provide more light and wider views.

Bead Molding: A small, cylindrical molding enriched with ornaments, resembling a string of beads.

Beam: A horizontal, load-bearing element that forms a principal part of a structure, usually using timber, steel, or concrete.

Blind Mortise and Tenon Joint: A joint in which the mortise does not extend through the stile.

Board and Batten: Vertical siding where wood strips (battens) hide the seams where other boards are joined.

Bottom Rail: The lower rail of the bottom sash of a double-hung window.

Bridging Solid: Wooden blocks used to separate floor joists beneath partition walls.

C

Capital: The elaboration at the top of a column, pillar, pier, or pilaster.

Casement Window: A window that opens by swinging inward or outward, much like a door. Casement windows are usually vertical in shape but often are grouped in bands.

Casing: The trim bordering the inside or outside of a window or door, commonly referred to as "inside" or "outside" casing.

Chair Rail: A wooden molding placed along the lower part of the wall to prevent chairs, when pushed back, from damaging the wall. Also used as decoration.

Chamfer: A beveled edge.

Chamfered Door: A door with a chamfer, usually at 45 degrees, that is notched or molded into the stile and rail. If the chamfer is limited in distance, it is called a "stop chamfer."

Classical: Refers to the architecture and design ideas of ancient Rome and Greece.

Clerestory: The fenestrated part of a building that rises above the roofs of the other parts. Upper elements of a Romanesque or Gothic church, bringing light into the center of the building from side windows pierced through stone.

Clerestory Window: A window (usually narrow) placed in the upper walls of a room, usually at an angle, to provide extra light.

Colonnade: A row of columns forming an element of an architectural composition, carrying either a flat-topped entablature or a row of arches.

Column: A slender, upright structure, usually a supporting member in a building. Freestanding or self-supporting structural element carrying forces mainly in compression; either stone, steel, or brick, or more recently, concrete.

Closed stair: A staircase enclosed on both sides by full-height walls.

Cornice: Decorative projection along the top of a wall; the upper part of the entablature from which the modern crown molding is derived.

Corinthian Column: In classical architecture, a column decorated at the top with a mixed bag of curlicues, scrolls, and other lavish ornamentation.

Corinthian: The type of Greek column characterized by simulated acanthus leaves.

Cornice: The uppermost section of moldings along the top of a wall; any molded projection of similar form.

Crown Molding: A molding in which the wall and ceiling meet; uppermost molding along furniture or cabinetry.

Dado: The zone between a chair rail or lower part of a sill and the baseboard.

Dentil: A small, square shape often repeated in a horizontal line.

Doric: The simplest of the three classical orders of Greek architecture.

Double-Hung Window: A window that operates by means of two sashes that slide vertically past each other.

Entablature: Made up of the architrave, frieze, and cornice, the entablature is the horizontal member that rests on the column and supports the roof.

Eyebrow Window: A small, horizontally rectangular window, often located on the uppermost story, aligned with windows below.

Fanlight: A semicircular or semi-elliptical window with a horizontal sill often above a door.

Fillet: The thin piece of wood used as a spacer between the tops of balusters.

Finial: A knob-like ornament.

Frieze: A band with designs or carvings along a wall, above doorways and windows, or on a mantel front.

Header Block: A Victorian trim detail used at the tops of doors and windows that was often quite ornate and carved. It sat on the corner like a rosette or a corner block.

Ionic: The type of Greek column characterized by scroll-like decorations.

Jamb: The vertical members of a window or doorframe.

Keystone: The central, topmost stone of an arch.

Meeting rail: The upper rail on the lower sash and the lower rail on the upper sash on a double-hung window in the closed position. It could be either a plain (meeting) rail or a check (meeting) rail.

Millwork: Finished woodwork, cabinetry, carving, and so forth.

Modillion: A bracket supporting the upper part of a composite or Corinthian cornice.

Molding: Shaped decorative outlines on projecting cornices and members in wood and stone.

Mortise and Tenon Joint: A joint in which a tenon is cut into the rail (horizontal piece) and inserted into a mortise on a stile. The mortise (often before 1890) extended through the stile. This strong joint was common from 1870 to 1890. Still used by some manufacturers today, it was replaced by the dowel joint.

Mullion: The vertical member separating adjacent windows.

N

Newel: The staring or terminating vertical piece that supports the handrail. Intermediary newels might exist wherever there are turns or ends during a stair run.

Niche: A recess in a wall to place various decorations.

Nosing: The rounded fore-edge of a stair tread.

O

Ogee Lugs: The stile extension beyond the check or meeting rail that is usually shaped or molded on the inside with an ogee profile. Also called "horns."

Oriel: A box-like window projecting from the wall of a house.

P

Palladian: A motif having three openings, the center one being arched and larger than the other two.

Palladian Window: A three-part window featuring a large, arched center and flanking rectangular sidelights.

Paneling: The lining of a wall with a wainscot.

Panel Mold: A separate trim piece nailed to a door to decorate or give a bolder appearance. It extends above the face of the stile and rail in a pronounced manner.

Pediment: A low, triangular gable above a cornice topped by raking cornices and ornamented. Used over doors, windows, or porches. A classical style.

Pendant: A bulbous, knob-like ornament that hangs downward.

Picture Window: One single, large windowpane that does not open from either side.

Pilaster: A rectangular, vertical member projecting only slightly from a wall, with a base and capital as well as a column.

Pier: A vertical, noncircular masonry support, more massive than a column.

Pillar: Similar to but more slender than a pier, also noncircular.

Plancher: Same as a bed board or soffit.

Plaster: A surface covering for walls and ceilings applied wet; dries to a smooth, hard, protective surface.

Plaster Board: A name applied to many commercial products on the market used as a backing for plaster.

Plate: The 2" x 4" (5.1 x 10.2 cm) trim nailed along the top edge of all stud walls. A plate also is secured to the top of all solid brick or masonry walls.

Plinth: The lower flat region of the base or pedestal in classical architecture.

Plinth Block: The large member used at the lower corners of doors where the base dies into its sides, and the casing dies into its top.

Ply Cap: A plain-shaped molding, rounded to provide a smooth edge along the baseboard.

Pocket Door: A door that slides open into cavities within walls, seeming to disappear when open.

Portiere: A curtain or grill work hung in front of on opening in place of or in conjunction with a door or a window.

R

Rafter: A roof beam sloping from the ridge to the wall. In most houses, rafters are visible from the attic. In styles such as a Craftsman bungalow and some "rustic" contemporaries, they are exposed.

Railing: A long strip of material running parallel to the stairs for use as a handhold while climbing or descending the stairs. Railings are typically supported by metal or wooden balusters and newels.

Raking Cornice: The sloping moldings of a pediment.

Riser: The vertical portion of a step. The board covering the open space between stair treads.

Rough Opening: The framed wall opening ready to receive a door or window unit.

Rough Sill: The bottom rail of a window's rough opening.

Run: The horizontal distance that a stair covers.

S

Sash: An individual window unit (comprised of rails, stiles, lites, and muntins) that fits inside the window frame.

Shutter: A movable cover for a window used for protection from weather and intruders.

Sidelights: Windows on either side of a door.

Sill: A horizontal piece forming the bottom frame of a window or door opening.

Skirtboards: Decorative lumber that covers stringers, often finished with carvings or molding.

Spiral Stair: A space-efficient stair design in which a helix of stairs rises around a central column.

Staircase: The complete assembly, including stairs, balustrade, and surrounding walls.

Stairwell: The enclosure of a stairway. The volume that a staircase occupies.

Sticking: The mold or profile on a stile or rail of a window or door.

Stile: The vertical sides of a window sash.

Stringers: The wide framing members that carry the treads of a staircase. Typically, stringers are made of lumber notched in a stair-step pattern to support the treads.

Stool: The inside windowsill.

Sunk Mold: A separate trim piece nailed to the panel on a stile and rail door that sit level with or below the face of the stile and rail.

Surround(s): The molding that outlines an object or opening.

Switchback: A stair arrangement in which the second flight turns 180 degrees from the first flight, typically at a landing.

Symmetrical: When two halves of an object are mirror images of each other.

T

Teram: The scroll at the end of a step.

Term: A pedestal resembling an inverted obelisk supporting a bust or merging with a bust.

Tread: The horizontal member on a staircase; the part on which one walks.

Tongue and Groove: A type of wooden siding with the edge of one board fitting into the groove of the next.

Top Rail: The upper rail of the top sash of a double-hung window.

Transom: A small window just above a door.

Tread: The horizontal portion of a step, usually with a rounded edge or "nosing" that overhangs the riser.

Turret: A small tower, often at the corner of a building. Common in Queen Anne styles among others. A turret is a smaller structure while a tower begins at ground level.

V

Veneer: A thin facing of finishing material.

Veneer Wall: The covering of one wall construction by a second material to enhance wall beauty. (Brick or stone over frame, brick or stone over concrete block.)

Vent Stack: A metal, plastic, or composite pipe (usually 4" [10.2 cm] in diameter) leading from the sewage network out through the roof to prevent pressure during sewage flow.

Volute: A common, decorative balustrade treatment in which the handrail makes a spiral curve away from the stairs before terminating at the starting newel.

W

Wainscot: A paneling applied to the lower portion of a wall.

Waste Pipe: The name generally applied to all household drainage pipes.

Water Closet: A commode.

Weather Stripping: A strip of fabric, plastic, rubber, or metal found around exterior wall openings to reduce infiltration.

Well Opening: A stair enclosure.

Winder: A series of turning steps in an L-shaped or switchback staircase. Typically, winders are used instead of landings in cases where space is critical.

Resources

A & M Victorian Decorations, Inc.
Manufacturer of architectural elements: mantels, columns, moldings, balustrades, wall caps, pavers, quoins, planters, urns, fountains, and gazebos; gypsum and cast stone; custom designs and finishes.
Phone: 800-671-0693; 626-575-0693
www.aandmvictorian.com

Adams Architectural Wood Products
Manufacturer of traditional custom wood doors, window sash and units, screens, storms, and combinations: all shapes, sizes, and wood species; glazing, restoration glass, mortise-and-tenon construction, true-divided lites, and more.
Phone: 888-285-8120

Advent Design International
Creator and supplier of hand-carved stone bathtubs, basins and vanities: stock and custom.
Phone 201-444-0426

Agrell Architectural Carving Ltd.
Custom fabricator of woodcarvings: hand-carved decorative moldings, capitals, brackets, furnishings, inlays and mantels; large-scale capacity for residential and religious buildings throughout the United States and Europe.
Phone: 415-381-9474
www.agrellcarving.com

American Wood Column Corp.
Manufacturer of custom woodturnings to match originals: non-load-bearing columns and balusters; fluted, plain and twisted; porch parts and newel posts; finials in any size and specified wood; clapboard siding.
Phone: 718-782-3163
www.americanwoodcolumn.com

Architectural Building Products by First Class
A widely diversified architectural products company: Columns, moldings, flexible moldings, ceiling domes, and ornament, balustrade, porch parts, cornices, architectural millwork in polyurethane, fiberglass, GRG gypsum, GFRC concrete, and other materials. Produces both standard- and custom-tailored products.
Phone: 770-514-8141
www.firstclassbp.com

Architectural Columns and Balustrades by Melton Classics
Manufacturer of architectural elements: columns, balustrades, moldings, cornices, porch parts, brackets, trim, and more; fiberglass, polyurethane, marble/resin composite, wood, cast, and synthetic stone; GRC and GRG.
Phone: 800-963-3060
www.meltonclassics.com

Architectural Columns by Arch Net Building Pro
Architectural columns and capitals available in a variety of styles and Orders of Architecture: Round, square, smooth, fluted, tapered, straight shaft, load bearing, and decorative. Offer columns made of fiberglass, fiberglass-marble composites, GRG gypsum, GFRC concrete, polyurethane, and a variety of wood species.
Phone: 770-514-8141
www.archcolumns.com

Architectural Components, Inc.
Custom fabricator of wood windows and doors: traditional details, materials and joinery; paneled, carved, louvered, French, pocket and art-glass doors; complete entryways; screen and storm doors; casings and moldings; mantels; replications.
Phone: 413-367-9441
www.architecturalcomponentsinc.com

Architectural Millwork by Polymouldings
Offers crown moldings, cornices, flexible moldings, chair rails, panel moldings, and framing trim for exterior and interior applications. Offered in long-lasting polyurethane, GRG gypsum, fiberglass, GFRC concrete, and flexible polymer resins; exceptional detail.
Phone: 770-514-8141
www.archcolumns.com

Architectural Pottery
Importer of Italian hand-carved sandstone and English-style cast stone: handmade Italian and Greek pottery; garden sculpture, planters and urns, fountains and benches; fiber-crete (FGRC).
Phone: 888-ARCH-POT
www.archpot.com

Architectural Products by Outwater, LLC
Manufacturer of 40,000-plus decorative building products: architectural moldings and millwork, columns and capitals, wrought-iron components, balustrading, door hardware, lighting, ceiling tile, furniture and cabinet components, and more.
Phone: 800-631-8375
www.outwater.com

Architectural Reproductions by Timeless
Designer and manufacturer of classically styled balustrade systems and decorative capitals and columns: lightweight, insect- and weather-resistant cultured marble, fiberglass and resin; smooth and detailed surfaces; CAD drawings and AIS specs.
Phone: 800-665-4341; 770-205-1446
www.timelessarchitectural.com

Architectural Stone Masonry
Manufacturer of custom natural-stone products using limestone and sandstone from North America and Europe: mantels, stairways, balustrades, columns, exterior wall cladding, archways, corbels and wall copings; design services.
Phone: 604-852-9662
www.stonemasonsarch.com

A Guild of Artisans
Founded in 1982, Artistic License is a group of dedicated professional artisans whose work continues time-honored traditions of crafts for the built environment. Members adhere to the highest stan-

dards of work quality for period architecture, restoration, interiors, and the decorative arts. See website for details about individual members. Many members work nationally.
Phone: 415-922-7444
www.artisticlicense.org

AZEK Trimboards
Manufacturer of cellular-PVC trim with the look of lumber: standard trim dimensions, sheet sizes and bead board; cut, routed, milled, and fastened using traditional woodworking tools; easily bent or shaped; exterior millwork.
Phone: 877-275-2935
www.azek.com

Bella Dura Architectural Stone
Fabricator of carved-stone elements: columns, capitals, balustrades, fountains, hand-carved mantels, and more.
Phone: 800-287-8976
www.belladura.com

Bendix Architectural Products, Inc.
Manufacturer of carved and embossed decorative wood moldings: rope, beaded, egg and dart, Greek key and fluted; plain, panel, and crown moldings; embossed wood ornament in ramin and oak; hand-carved capitals, corbels, onlays, mantels, and more.
Phone: 201-567-1003
www.bendixarchitectural.com

Camcraft 3D, Inc.
Custom wood carver: computer-controlled design and manufacturing; deep-sculpt moldings, panels and curved and carved casing; machine carvings, architectural ornament, sculpture, mantels, capitals, carved doors, and signage; all wood species.
Phone: 713-550-8544
www.wooddesigner.com

Cantera Especial
Manufacturer of hand-carved natural-stone products made from limestone, cantera, adoquin, travertine, marble, and sandstone quarried in Europe and Mexico: fireplaces, fountains, columns, balustrades, molding, sculpture, and custom work.
Phone: 800-564-8608
www.cantera-especial.com

Chadsworth's 1.800.Columns
Manufacturer of authentic interior and exterior molded ornaments: columns and capitals, pergolas, pillars, pilasters and balustrade systems; plain or fluted shafts; molded polymers, polymer/stone composites, and wood.
Phone: 800-265-8667
www.columns.com

CinderWhit & Co.
Manufacturer of historically and architecturally correct exterior and interior wood turnings: porch posts, balusters, newel posts, finials, and spindles; stock or replica/custom designs; handrails.
Phone: 800-527-9064
www.cinderwhit.com

Craftsmen Group, Inc.
Traditional woodworking for historic buildings, restorations, reproductions, glazing, traditional finishes, hardware, installation, doors, and windows.
Phone: 202-332-3700
www.thecraftsmengroup.com

Crown Point Cabinetry
Hand-crafted period-style cabinetry of the finest quality.
Phone: 800-999-4994
www.crown-point.com

Cumberland Woodcraft Co.
Manufacturer and distributor of handcrafted millwork and ornament: Spanish cedar, poplar, and mahogany; licensee of the Victorian Society in America for wood trim; columns, screens and storms, trim, balustrades, cornices and porch parts.
Phone: 717-243-0063
www.cumberlandwoodcraft.com

Decorators Supply Corp.
Supplier of 14,000 patterns for period architectural elements and molded ornaments: cornices, columns, capitals, mantels, ornamental ceilings, niches, domes, brackets, and corbels; plaster of Paris, wood, and compo; since 1893.
Phone: 773-847-6300
www.decoratorssupply.com

Dixie-Pacific Manufacturing Corp.
Manufacturer of architectural components: fiberglass and wood columns and decorative capitals; synthetic balustrades, porch posts, and stair parts; QuickRail, a synthetic railing system.
Phone: 800-468-5993
www.dixie-pacific.com

DJ Studios
Custom fabricator of molded ornament: ceilings, columns, mantels, and more; Forton, GFRC, plaster, and polystyrene foam; bonded metals and metallic finishes.
Phone: 770-798-9075
www.djstudios.net

Erik Wyckoff Artworks
Custom designer and fabricator of hand-carved architectural woodwork: original entry, interior and wine-cellar doors; most wood species.
Phone: 612-617-0446

Fagan Design & Fabrication, Inc.
Manufacturer of architectural elements: wood columns, cylinders, rope twists, large turnings, octagons, and pilasters; Classical order; load bearing and ornamental; columns also in plaster/gypsum, custom FRP and GFRC.
Phone: 203-937-1874
www.fagancolumns.net

Focal Point Architectural Products
Manufacturer of high-density polymer moldings, medallions, niches, and fiberglass columns; Class A fire rating on some products.
Phone: 800-662-5550
www.focalpointtap.com

Forshaw of St. Louis
Custom fabricator of mantels: cast stone and plaster; pine, oak, poplar, cherry, and other solid hardwoods; precast mantels for 33-, 36-, 42-, and 43-in. (83.8-, 91.4-, 106.7-, and 109.2-cm) openings; wood mantels fit any size fireplace; stone mantels fit 36- to 42-in. (91.4- to 106.7-cm) fireplaces.
Phone: 800-367-7429
www.forshaws.com

Foster Reeve & Associates, Inc.
Custom fabricator of fine ornamental and architectural plaster details, specialty plaster wall finishes and stock moldings: design development, engineering and plaster program-management services.
Phone: 718-609-0090
www.fraplaster.com

Fypon, Ltd.
Manufacturer of more than 4,000 molded architectural elements: window features, balustrades, turnings, ceilings, brackets, and more; high-density polymer and other polymers; load-bearing polymer/steel columns, 12-in. (30.5-cm)-maximum diameter.
Phone: 800-446.3040
www.fypon.com

Gingerbread Man
Manufacturer of Victorian gingerbread, screen doors, porch parts, balustrades, gable ornament, ceiling medallions, fence pickets, sawn balusters, brackets, signs, roof ridge cresting, cupolas, arbors, and more: custom design services.
Phone: 530-622-0550
www.gingerbreadman.com

Goodwin Associates
Supplier of interior and exterior architectural building products: columns and capitals, balustrade systems, moldings, fireplace surrounds, domes, medallions, metal ceilings, and more; polyurethane, wood and fiberglass; stock and custom.
Phone: 585-248-3320
www.goodwinassociates.com

Goodwin Heart Pine Company
Manufacturer of antique river-recovered heart pine and heart cypress reclaimed from Southern rivers: for flooring, stair parts, furniture, and moldings; building-reclaimed wood; custom orders; 15 grades.
Phone: 800-336-3118
www.heartpine.com

Haddonstone (USA), Ltd.
British- and U.S.-based manufacturer of landscape ornament and architectural cast stonework: balustrades, columns, capitals, porticoes, cornices, molding, trim, molded panels, and more; fountains; custom components; 200-page catalog.
Phone: 856-931-7011
www.haddonstone.com

HB & G
Manufacturer of PermaPorch system: load-bearing PermaCast columns, PermaWrap columns, grand balustrade systems, PermaPorch and deck railings, load-bearing PermaPost porch posts, PermaCeiling, moldings, medallions, and entrance features.
Phone: 800-264-4424
www.hbgcolumns.com

Historic Doors
Custom manufacturer of wood windows and doors: circular casings and moldings; restoration and period-style construction; garage doors.
Phone: 610-756-6187
www.historicdoors.com

Historical Arts & Casting, Inc.
Designer and custom fabricator of ornamental metalwork: columns, lighting, grilles, doors, windows, kitchen hoods, and more; cast iron, bronze, aluminum, and wrought iron/steel; Arts and Crafts, Victorian and other styles; restoration services.
Phone: 800-225-1414
www.historicalarts.com

JMS Wood Products
Supplier of rope moldings from 3/8 to 3 inches (1 to 7.6 cm) in diameter and rope columns from 4 to 24 inches (10.2 to 61 cm). in diameter: rope, fluted and twisted designs for stairs; plinth blocks for door surrounds; porch parts; mantels; any wood species.
Phone: 818-348-7230
www.jmswoodproducts.com

Josef's Art Wood Turning
Manufacturer of interior and exterior columns: fluted, reeded, roped, and carved; posts, balusters, railings, table and chair legs and other turnings; any wood species; in-stock inventory and custom work.
Phone: 516-489-3080
www.jawsinc.com

Klitsas, Dimitrios – Fine Wood Sculptor
Custom sculptor and carver of wood architectural elements: interior and exterior; furniture in all period styles, capitals, mantels, moldings, and specialty carvings.
Phone: 413-566-5301
www.klitsas.com

Knickerbocker Studio
Wood Carving, no machines; all hand carved, interior molded ornament, columns and carved capitals, interior woodwork, carved paneling.
Phone: 207-633-3818
www.knickerbockergroup.com

Limestone Concept, Inc.
Custom fabricator and distributor of hand-carved elements: fountains, columns, balustrades, urns, benches, pavers, and statuary; antique mantels; French limestone slabs, and tile; antique terra cotta, flooring, and quarry tile.
Phone: 310-278-9829
www.limestoneconcept.com

Manor Style, Ltd.
Nationwide supplier of architectural elements: custom fabricator of components in DuraStyle and AZEK PVC; ceiling medallions, paneling, balustrades, cornices, mantels, molding, and trim; load-bearing fiberglass columns and capitals.
Phone: 800-325-2188

Maurer & Shepherd, Joyners
Manufacturer of historic reproduction architectural millwork: windows, doors, entryways, and raised paneling; pine and mahogany.
Phone: 860-633-2383

Michael A. Dow, Woodcarver
Custom hand wood carver: architectural, ornamental, and nautical elements; furniture, mantels, molding, capitals, and signage; wood turnings; any style, and wood; model making and antique carving restoration.
Phone: 207-363-7924
www.archcarving.com

Old Wood Workshop
Supplier of antique reclaimed and resawn wood flooring and vintage building materials: hand-hewn beams, antique doors and period iron hardware; manufacturer of custom tables and countertops in antique wood; mantels.
Phone: 860-655-5259
www.oldwoodworkshop.com

Olde Good Things
Architectural-salvage firm: antique wood flooring; period hardware, doors, mantels, columns, moldings, lighting, furniture, sinks, faucets, tile fireplace surrounds, fencing, and garden ornaments; terra-cotta, support beams, and rafters.
Phone: 212-989-8401
www.oldegoodthings.com

Southern Group Enterprise, Inc., Unique Mantel Co.
Manufacturer and supplier of hand-carved marble mantels: all styles; stock designs, and custom work; fountains and statues in stock.
Phone: 909-464-1818
www.uniquemantel.com

Superior Moulding, Inc.
Supplier of standard and custom moldings and more: embossed, sculpted, and polyfoam moldings; columns, capitals, ceiling medallions, niches, domes, corbels, furniture legs, windows, doors, stair parts, hardwood flooring, and more.
Phone: 207-625-7000
www.sunarchitecturalwood.com

The Wood Factory
Manufacturer of historically correct Victorian millwork: interior and exterior doors and stair parts; screen doors, porch posts, newel posts, rails, balustrades, brackets, capitals, custom mantels, siding, garden specialties, and more.
Phone: 936-825-7233
www.thewoodfactory.com

Vintage Woodworks
Supplier of Victorian millwork: western red cedar shingles, porch parts, columns, turned and sawn balusters, railings, brackets, gazebos, cornices, corbels, spandrels, mantels, storm and screen doors, and more.
Phone: 903-356-2158
www.vintagewoodworks.com

White River Hardwoods–Woodworks, Inc.
Manufacturer of architectural millwork: Mon Reale moldings, authentic hand-carved linden products, and adornments for cabinetry and furniture in cherry and maple; in-stock lineals, finials, mantels, and range hoods.
Phone: 800-558-0119
www.mouldings.com

Wilbur, Frederick–Woodcarver
Wood carver: traditional decorative interior and exterior carvings; furniture, mantels, moldings, friezes, capitals, rosettes and heraldry; original designs and historically accurate reproductions.
Phone: 434-263-4827
www.fredrickwilbur-woodcarver.com

WindsorOne
Manufacturer of historically inspired Moldings Collection: whole-room-style concept in Greek Revival, Classical, Craftsman, and Colonial Revival; engineered wood board siding, end-and-edge glued, finger jointed, and double primed.
Phone: 888-229-7900
http://www.windsorone.com

Woodline Co.
Manufacturer of hand- and machine-carved architectural wood elements: corbels, capitals, columns, balusters, newel posts, rosettes, and more; many species; stock and custom designs.
Phone: 562-436-3771
www.woodlineusa.com

Yarrow Sash & Door Co.
Custom designer and manufacturer of handcrafted wood doors and windows: historic replication; mortise-and-tenon joinery; all shapes and sizes; entry doors and screen and storm doors; in-swing French casement and double-hung windows; many species.
Phone: 877-237-8650
www.yarrow.mb.ca

Directory of Designers

Bernhard & Priestley Architecture
www.bp-architecture.com
Page 136

Richard Burt Architects
www.richardburtarchitects.com
Page 142

John Cole
www.johncolearchitect.com
Page 130

Crown Point Cabinetry
www.crown-point.com
Page 134

Fairfax & Sammons
www.fairfaxandsammons.com
Page 96

Krumdieck A +1
Page 103

Christine G. H. Franck
www.christinefranck.com
Page 94; 108; 120; 140

Fanny Haim & Associates, Inc.
www.fannyhaim.com
Page 112

Historical Concepts
www.historicalconcepts.com
Page 106; 118; 124

Hutker Architects Inc
www.hutkerarchitects.com
Page 146

John B. Murray Architect
www.jbmarchitect.com
Page 104

Elliott Elliott Norelius
www.elliottelliottnorelius.com
Page 122

Benjamin Nutter Associates
www.benjaminnutter.com
Page 92; 132

Polhemus Savery DaSilva Architects Builders
www.psdad.com
Page 144

Rill & Decker Architects
www.rilldecker.com
Page 128

Eric Watson, Architect
www.ericwatson.com
Page 148

About the Author

For sixteen years, Nancy E. Berry has been writing on the subject of interior design and architecture as a staff editor for magazines such as *Atlanta Homes and Lifestyles*, *Cape Cod Home*, and *Old-House Journal*. Additionally, Berry hosted the weekly radio show "Atlanta Homes and Lifestyles Live" on the topic of design. Today, she is the editor of *Old-House Journal's* sister publication *New Old House*.

Acknowledgments

As I was growing up in an old Victorian in Newton, Massachusetts, my home had seen its fair share of remuddling over the years. Thanks to my father, Frank Downey, for involving his children in all the home improvement projects. He taught me how to use a hammer and, although at the time I was most ungrateful, his tutelage has been invaluable in my own home projects. I would also like to thank Brent Hull of Hull Historical and Jeffrey Davis of Chadsworth's 1.800.Columns for providing sound advice and great tips for the book. I would also like to thank all the architects and designers who offered their expertise to the Design Details sections. Special thanks to my friends and family—in particular, Richard Downey—who shared their trials and tribulations of incorporating trim into their own home projects. A very special thanks to illustrator Rob Leanna for all his beautiful artwork. And thank you to Steve Thomas and Woodmeister for location assistance.

Photographer Credits

Sandy Agrafiotis
72

Gordon Beall
6; 11; 12; 13; 26; 36; 37; 55; 88

Creative Publishing International
7; 79; 82; 85; 87

Carlos Domenech
113

Courtesy of ICI Paints/ www.icipaints.com
4 (bottom right); 17

Erik Johnson/Chadsworth's 1.800.Columns
28; 29; 57; 95; 109; 121; 141

Richard Leo Johnson/Atlantic Archives
19; 30; 47; 107; 119; 125

Erik Kvalsvik
103

Shelley Metcalf
51; 69; 115

Rill and Decker
4 (bottom left); 129

Paul Rocheleau
43

Eric Roth
22; 40; 59; 61; 78; 84; 93; 127; 133; 135; 139

Richard Sexton
149

Robin Stubbert
75; 86; 91; 99; 117

Linda Svendsen
49

Brian Vanden Brink
27; 53; 71; 81

Brian Vanden Brink/Bernhard & Priestley Architecture
65; 137

Brian Vanden Brink/Rick Burt, Architect
143

Brian Vanden Brink/Catalano Architects
8

Brian Vanden Brink/Centerbrook Architects
101

Brian Vanden Brink/John Cole, Architect
131

Brian Vanden Brink/Elliott Elliott Norelius Architects,
123

Brian Vanden Brink/Hutker Architects, Inc.
147

Brian Vanden Brink/Dominic Mercadante, Architect
21

Brian Vanden Brink/Doreve Nicholaeff, Architect
39

Brian Vanden Brink/Polhemus Savery DaSilva Architects
145

Brian Vanden Brink/Rockport Post & Beam
4 (top)

Brian Vanden Brink/Winton Scott Architects
111

Jonathan Wallen
45; 97; 105

James Yochum
15